NUNC COGNOSCO EX PARTE

TRENT UNIVERSITY
LIBRARY

The Liberal Future

THE LIBERAL FUTURE

by

J. GRIMOND

FABER AND FABER
24 Russell Square
London

*First published in April mcmlix
by Faber and Faber Limited
24 Russell Square London W.C.1
Second impression May mcmlix
Printed in Great Britain by
Latimer Trend & Co Ltd Plymouth
All rights reserved*

*© Joseph Grimond
1959*

Contents

Author's Foreword	page	9
1. Liberal Political Principles		11
2. Liberal Political Theory		24
3. Liberal Political Reforms		36
4. Liberal Economics		54
5. Co-ownership		79
6. The Social Services		98
7. Education		113
8. The Liberal Environment		131
9. British Liberalism and the Rest of the World		151
10. Epilogue		180
Appendix		184
Index		195

Foreword

Anyone who expects to find in this book a list of measures which a Liberal Government in office would propose will be disappointed. I have not tried to present a political programme, though in the Appendix I have outlined some aims which Liberals should pursue. What I have tried to do is to put down my thoughts on those larger issues which must dominate all detailed policies. No reader of any party can be expected to agree with everything in this book; indeed, I claim freedom for myself to change my mind on minor points. I shall have succeeded if I have drawn a sketch of a Liberal society—and it must be looked at as a sketch, not a detailed picture. I have produced my sketch in some four months during which I have been leading a fairly busy life in other ways. Could I have had the time to chisel and polish every sentence it might have been better—but, equally, it might never have been written at all.

Everyone will be able to point to omissions. I have not, for instance, dealt at any length with the protection of the individual against bureaucracy. My reason for this and for some similar omissions is that I believe we should examine causes rather than effects. Why is the individual threatened? If you have a wrong distribution of power in a society, you should try to cure the fault, not merely apply piece-meal remedies and palliatives. Further, on this particular subject of individual rights there is no better solution than the simple one of ready recourse to the courts of the land.

The disregard for private rights of which we have had some instances in recent years is serious, but there are, to my mind, more potent evils still in politics today. There is, for instance, our failure to seize on the chance which science is offering us of getting away from the dreary caricature of a human race moved entirely

FOREWORD

by economic motives: the Marxists and the early Liberal economists may have been right in their day to stress the immediate economic needs of Britain; even now no one can be complacent about the production of sufficient wealth or the standard of living in Asia and Africa; but in Europe we must rise above the class war. If we could reject war between rich and poor, between owners and Trades Unionists, as the basis of our politics, we could be free to give more attention to supplying economic and political aid to those countries in need of it, as well as to other constructive endeavours.

But we cannot do this until we have reasserted our own values. The values which democracy serves are more important than democracy itself. In the West, however, there are signs of cynicism about values and of an exhaustion thanks to which the difficulties loom larger than the possible prizes. There is an undercurrent of violence; there are those who demand that someone else should give them a lead—the classic cry of cowards in all ages; there is the growing band of those Pilates who brilliantly criticize our politics but wash their hands of any involvement in politics. No doubt there have always been sluggards and doubters: but they can do particular damage today when liberal democrats are living in a rapidly expanding Universe on which they are being jostled every day by new thoughts, new techniques and beset by the growing menace of Russia and China.

Many people have helped me with this book. They may well prefer not to be implicated in the result, but I must in particular thank Arthur Holt, M.P., Dr. Nathaniel Micklem, Mark Bonham Carter, Mr. Roger Fulford and Frank Ware.

1

Liberal Political Principles

Liberalism is a personal matter. It is about individual political conduct and individual freedom. Liberals see no salvation which is not personal. They recognize no values which are not ultimately personal values. If man in Liberal eyes has a strong taint of original sin he is eminently redeemable. Society, while curbing his sinful impulses, should give him a chance to work out his political salvation for himself.

The purpose of political Liberalism is to create conditions in which personal Liberal principles can be practised. The processes of Liberal government are reciprocal; a Liberal government must both be run on Liberal principles and promote Liberal principles. Liberal democracy is more than the casting of a vote. It is more than the success of certain political practices or institutions. It is the acceptance and practice of these principles by individuals. I say acceptance and practice because unless there is understanding or acceptance of these principles their practice will always be precarious: but if there is only acceptance without practice (as in many ways there is today) Liberal democracy dies.

I cannot give a complete list of the principles or qualities of Liberalism, much less put them in order of importance. They are familiar in a vague way to many people in this country, though by no means to all. Lately they have been confused. Imprecise adjectives have been attached to fairly precise nouns as in the phrase 'Social Justice'. Some of these principles have been sneered at. 'What is the good of freedom to starve?' we are asked. Many of these Liberal qualities mean very little for most people: they arouse little enthusiasm, and their violation little protest. So some restatement of Liberal qualities, however inadequate, must start this book.

LIBERAL POLITICAL PRINCIPLES

High on the list I would put responsibility. We, as individuals, must be prepared to take responsibility for what we do or advocate. The notion that it is not my business when wicked things are done or matters within my reach go wrong is wholly illiberal. It led to toleration of anti-Semitism and the concentration camps in Germany. It may yet break French democracy. Curiosity, questioning, criticism are essential to Liberalism. Liberals assume change: but they do not assume that its direction is inevitable. They are determined to use and mould change. They assume that by using their intelligence and their moral sense men can make themselves better off materially and morally. But this process demands continual criticism and argument. In political life one of the first principles is to examine all dogma and all institutions. Dogma and institutions will always be with us: but it is illiberal, for instance, to curtail the rights of Communists or atheists to state their beliefs.

There should be no limit on criticism, even of our most hallowed beliefs. Of course there should be laws of Treason, Libel and Criminal Libel: but these are general rules aimed at certain actions. There should be no attempt to put the Crown, the established religion, or the British Way of Life for that matter, in a privileged position above examination and dissent. Once you put any institution or belief on a pinnacle of its own, you open the way to censorship of all kinds and you end by killing the thing to be protected, for lack of criticism will destroy it.

Reliance on reason is also a Liberal virtue, and much needed in a Liberal Society. Many may say that today it is sufficiently accepted that we should try to settle our differences by rational argument. They will feel that it is as intolerably presumptuous for Liberals to lay claim to reason as it is for Tories to annex patriotism. But look round and everywhere you see reason in the gutter. It is not only in Germany, Italy and Russia, that unreason has triumphed in our time. Suez brought to light unreason in this country; McCarthy did so in America. Nationalism, racialism, anti-Semitism, emotion of all kinds remains potentially the strongest political motive—far stronger than reason unless reason constantly asserts itself as in this country under Tory and Labour Governments it has failed to do.

LIBERAL POLITICAL PRINCIPLES

dilemma. 'Ends' and 'Means', though separate, cannot be completely divorced. As human beings must act as though they had some moral choice, and must believe that the right choice can be recognized, they will find that Free Choice is always inextricably bound up with the Right Choice. If you believe that people might freely choose such things as the legalization of indiscriminate murder of one's children, then there is really no basis for Liberalism at all. Liberals, at any rate, believe that there are such things as right decisions and that men will usually reach them if they are given a fair and free chance to do so. But this does not mean that everyone will be right about every decision which he takes. The Right Choice, like other correct decisions, will often only be reached by checking one's feelings against other people's. The environment in which choices are made must be Liberal in the sense that it allows these choices not only to be free, but to be checked against other free choices—and checked before they are put into practice. There is no contradiction between freedom to choose and checking one's choice. In morals the right choice is no more often obvious than in other matters. We face competing claims and between these claims we need help to decide correctly. What is destructive is the ultimate submission of our will simply to what we are ordered, without any consideration of whether it is right or wrong. This is the situation in which power corrupts both the ruler and the ruled.

These conditions are essential to Liberalism: that people will try to exercise Liberal qualities, and that they are in a position to do so effectively. If they are absent, nothing, no tinkering with institutions, no laws or political systems can save Liberalism. If no one will tolerate opposition: if everyone is moved only by immediate self-interest: if the strongest is free to bully and the weakest have no elbow-room at all, Liberalism is doomed.

The conditions of good political life depend not on direct action by the Government, but on a good society. Unfortunately, 'society' and 'community' have repulsive overtones. The former tastes of the *Tatler*, starchiness and class distinctions: the latter of prigs and busybodies and professional Sociologists: both of a stuffy, precious and wholly illiberal world. But this is because they have been abused. 'Society', as I use it, means simply the

LIBERAL POLITICAL PRINCIPLES

variety of relationships which individuals feel agreeable and useful. If I live in a village, never go to church or join in village activities, yet if I feel myself part of the village, see a few people in my own home and generally accept the way of life, I say I live in its society.

At the root of a Liberal society lies a belief in the worth—leaving aside equality or inequality of the worth—of everyone and their right to go their own way.

'The Poorest He that is in England has his life to live as the Greatest He.'

This is one Liberal text. And it is more distinctive than may at first appear. It asserts the individual and the value of any individual—even the poorest He. But it asserts it without envy. It does not demand that the rich be made poor—nor even claim that the poor are more deserving than the rich. It demands equality in one thing only, the right to live one's life. That life must be lived in a Society: 'The Free Society', 'The Open Society'. These are phrases used by Liberals, but what exactly do they mean? As far as I am concerned, with all respect to Mr. Lippman, I intend like Humpty Dumpty to make them mean what I want them to mean.

To take the word 'Society' first. It emphasizes the individual, but the individual in conjunction with other men and women.

When the dancers arrive for a ball they are jumbled together like marbles in a bag. When the dancing starts, however, they form patterns and assume relationships to one another, which are both changing and voluntary. They become a society. The pattern of society is set by its friendships, customs, tradition, morals. There are distinctions in a society, as there are different roles for different players at football, but they are not wholly hierarchical. A society maintains its own organizations apart from government. It is superior to government. Government is its servant. A hallmark of a society is that it is bound together by a general morality which is interpreted in different ways in various institutions, but which rests on the broad pillars of a general agreement. The common law is a reflection of our general understanding of how affairs should be conducted. It is impartial, traditional, full of common-sense and intelligible. Churches, families, clubs, the Stock Exchange, the Trade Unions make their rules and enforce them by assent. They differ from but are compatible with our

LIBERAL POLITICAL PRINCIPLES

general morality. It is not a rule of the Stock Exchange that you may cheat, nor of a family that the strongest can knock the others about.

A Society should have its standards. Its aristocracy should be one of talents which is respected. Equality of Opportunity must lead to inequality in other ways: but individual position, power, privilege should only be pursued to the point at which they still benefit the community at large and do not harm it.

The opposite of a Society is a Proletariat. The proletarians are marbles in a bag. The proletarians maintain no institutions of their own. The proletarians are means, not ends. They are equal with the full horror of the equality of universal sameness. They are 'hands'—they are statistics. This is not to say that they may not be very well looked after. One of the illusions which Marxists have left is the illusion that a proletariat is necessarily poor and oppressed. A proletariat may be copiously supplied with cars, houses, television sets, and still be a proletariat, while a society may be primitive, but still be a society. A proletariat may be rich. America today has some symptoms of a proletarian state. The Americans on the whole are not wage-slaves. Nor do they live in fear of any secret police. They run a genuine democracy. They change and criticize and not only feel free but are free. And yet in so far as they have sacrificed all effort to reach individual perfection in exchange for an overall mediocrity, and when they submerge their identities to present a target for more and more dish washers and motor cars rained down on them by a heavier and heavier barrage of advertisements, they are a proletariat. They are in danger of losing control of their own destiny. They are, in fact, saved from being a proletariat by other features of their way of life. But the possibility is there and, indeed, the possibility is here.

The essence of a proletariat is that there are two groups, the rulers and the ruled. The rulers may be big business or the Government. Other group relationships are feeble. All the horizontal threads are weak. The mass of undifferentiated citizens are not bound together, they are bound to the ruler.

But not all societies are free. There are closed societies which are as illiberal as any proletariat.

A closed society is enforced. You cannot contract out of it in any way. You cannot even criticize it without risk. In such a society the hierarchy tends to be rigid whether it is an aristocracy of birth, or of position, or of wealth; its members claim privileges for their order as such, not for their merit. They resist outsiders. They are, like the tribe, anxious to perpetuate their order and its position.

Looseness, growth, change, freedom of movement, these are the marks of a free society. But not weakness. It may have strong and indeed compulsive institutions. But they must be open—open to receive new ideas, new members, new criticisms. Like all strong things, Society can be a danger. It used to horrify Conservatives, and indeed Liberals and Socialists, to hear tales of Communist children being brought up in communal crèches. But there are few things as tyrannical as the family. Yet the family is the father of all societies. The power of the Trade Unions frightens many people, who are also kept awake at night by fear of George Orwell's 1984. But the Trade Unions as Medium Big Brother may be a champion against the State as Very Big Brother. The point is that the organs of society may be strong but that they should be continually subject to criticism, that they should depend on assent and superiority of fitness, not on sanctions, and that there should be vigorous rebel movement within society against its established institutions. Membership must in the last resort be voluntary: though the claims of society can legitimately be very strong they must be weighed against other claims. Members of society retain control of their own destinies. They do not surrender them to society—still less to their government.

The dangers inherent in all societies have led all too often to a denigration of any society. If an all-powerful Church or aristocracy or plutocracy represses freedom, degrades standards and stifles the criticism upon which quality depends, then this is a bad society which needs to be revived by the emergence of nonconformists acting through or against its institutions. But the fact that standards may be wrong, that our relations with other men and women may be difficult, even restricting, does not mean that we should have no standards and that primitive freedom in its negative form is the highest form of community human beings can obtain.

LIBERAL POLITICAL PRINCIPLES

The Government should be one means by which we regulate society. 'Government is a contrivance of human wisdom to satisfy human want.' So wrote Burke, and gave us another Liberal text. Burke, like the Bible, is quoted by strange followers and used to lend pith to all sorts of queer doctrines. But this saying of his is a thoroughly Liberal saying. Again, it too, like my earlier quotations, may look flat to the point of platitude. But it is, in fact, a revolutionary view in the twentieth century. Germany almost deified Hitler's Government and the Communists have set their Governments up as idols. They have endowed them with the magical powers of heathen gods. The machinery of government has been veiled in awe and described as a monster father—just as were primitive gods.

Nor need we go to the extremes of tyranny to see this obsession with government. A paradox of the free world is that while respect for its politicians has worn rather thin, the inevitability, the utility, the inexorable swelling of the government machine is taken for granted.

To prune government back to its true task is assumed to be impossible. To lay more and more on the back of this animal, which we now all assume to be a rather weak and foolish animal, is assumed to be sense.

Liberals must consider how this situation has come about in which society has been so much weakened and Government so greatly extended. They must consider it all the more carefully because it holds the key to their failure and because they themselves are partly responsible for it. A Free Society is what Liberals want, yet they themselves weakened its fabric and let in the heresies which have almost destroyed it.

The Liberal Party performed wonders in the nineteenth century. They carried through the extension of the franchise, they created the conditions for a material advance, the like of which has never been seen. Nor did they neglect what we now call social legislation. Even before 1906 most social legislation was Liberal.

The industrial nineteenth century was not a scene of unrelieved exploitation and misery. I am glad that a more balanced view is now being taken of what is loosely called nineteenth-century *laissez-faire*. Adam Smith and his successors never preached economic civil war with the weakest always to the wall. Liberal thinkers and economists were deeply and effectively concerned with the condition of the people. But not effectively enough.

In foreign affairs Liberal policies freed Europe for decades from major wars. In the Empire Liberalism stood for humanity and progress. Most remarkable of all was the success of Gladstone in rousing the nation to its responsibilities. It is almost incredible today, when politics is so often a question of getting all you can, to read his speeches rousing his audiences to give all they could.

But the nineteenth century was too callous about poverty and with all its successes nineteenth-century Liberalism, in my eyes, failed in another vital respect. It watched, indeed it encouraged, the break-up of the old society without bringing about a new one. The apocalyptic forces of the industrial revolution tore men and women from their moorings and set them adrift without succour.

It all too often allowed them to be used as mere fodder for the industrial machine. The increasing population was herded into new or vastly expanded towns with too little regard to human values. Class was separated from class. Little trouble was taken to see that channels of democracy were open and working. Human beings met the nineteenth century with courage and ingenuity. They built, not only the massive numbers of houses, roads, railways and factories at which we are all too apt to sneer today. They also rebuilt their lives. They blossomed out in a multitude of churches and clubs and societies. They were willing to make sacrifices for education and to attempt the betterment of themselves and the country. But the disruption was there nevertheless, and Liberals did too little to heal the wounds. Liberal political philosophy declined in the last half of the nineteenth and the first half of the twentieth century. The impetus left by the eighteenth and early nineteenth centuries was so strong that we have hardly yet realized that the machine was running down. Even now, if you listen to any discussion about political advance in a colonial territory, it is usually conducted on the lines of a hundred years ago. There has been no advance in political thinking or techniques comparable to those in the scientific fields.

A need was there. Inhumanity was abroad in the factories and slums. In their search for a solution men turned to the State. It

LIBERAL POLITICAL PRINCIPLES

was called in to right the wrongs. The Marxists analysed the tendencies at work; analysed them forcefully but illiberally. Socialists of various sorts also looked to the State. The Liberal Government of 1904 to 1916 used the State. Indeed, it was inevitable that the State should be used. The Liberal Government was, in my opinion, right in founding the Welfare Services. But Liberals were wrong in not probing the root causes of the ills which the State was to cure. It is true that so long as the population outstripped the means of supporting the bulk of the people at a decent standard there was no instrument but the State which a Government could use to help those who fell below subsistence level. But Liberals who have always been jealous of too much government were curiously blind to the dangers around them. State Socialism in Germany was aggressive. In Britain and in France the forces of political and economic nationalism were growing. Yet Liberals took little trouble to define the proper limits of a central government. They were slow to see what a high standard of responsibility was required before you could entrust such powers to a government. They paid little attention to the inadequacy of the sources of information which were available to the growing electorate. They developed no sufficient enthusiasm for social welfare work, and they were heedless of the motives of the great body of the people. Above all, they lost sight of their main purpose, which should have been to give back power and choice to the people themselves, to reconstitute Society, not to enthrone the State.

So the field was left to Socialists. They held that voluntary effort was unable to cope with the situation. They were prepared to use the State even if it meant overriding personal liberty, even if it meant the treatment of the citizens as a proletariat. The subsequent history is there for all to read. The State attaching to itself all the emotions of nationalism became more and more powerful. The central government was extended into all sorts of spheres regardless of its suitability.

This was bad enough. National Socialism has been the main cause of a half-century of international anarchy, two world wars and the use of force as an instrument of foreign policy. But now the British Labour Party are finding that even those jobs inside the nation which Socialism claimed to be able to do have either been completed or are impossible. Nationalization has proved a fiasco. If it is pushed to its logical conclusion it is incompatible with freedom. So is the promise of endless welfare benefits to be handed out by the grandmother state. The British Labour Party has changed a great deal. Building Controls, Import Licences, the Purchase Tax, the Capital Issues Committee, all these expedients in which it now trusts were piled up during and after the last war when they were introduced to meet a situation of siege. They have little to do now with either Marxist or Christian Socialism, for they have no logic and no moral framework behind them.

But Socialists remain loyal to their belief that the State must be the engine of all advance. They called on the State to fill the deficiencies left by nineteenth-century Liberalism, and they have taught that salvation lies in entrusting it with more and more power. The Tories have accepted the new dispensation; they are now conservative of Socialism as once they were conservative of Liberalism. Liberals themselves have not been free from the prevailing opinion. Indeed, in the Liberal Yellow Book of the thirties they put forward proposals which demanded a great deal of action by the State. In some ways the thinking in that book is a coherent statement of the kind of policies which post-war Labour and Conservative Governments have been trying to pursue in a haphazard fashion. Liberals may claim that the tragedy lies in the failure of Conservative Governments of the thirties to put these policies into practice before the war when they were most apposite to the troubles which faced us. Since Keynes (who, it is sometimes forgotten, was a Liberal) and the Yellow Book, the Liberals cannot be described as a party of pure *laissez-faire*, which indeed they never were.

But it is my contention that the time has come to consider the true spheres of Society and the Nation-state so as to strengthen the former as a bond uniting all citizens and at once prune and improve the latter. Liberals are not opponents of all public intervention. They have very definite ideas, however, on how and by whom it should be done. But the key to planning is responsibility. Modern Governments, with their ostensible faith in conscious planning, have given very little thought to this. Under their hand

LIBERAL POLITICAL PRINCIPLES

so-called responsibility has been heaped on ministers and bodies who cannot and do not exercise responsibility. The planned State has become the irresponsible State. What we need is a Liberal Society with an effective State to serve it in the State's proper sphere.

But before I come to our political and economic proposals I must deal briefly with the Liberal attitude to one other human institution—the nation. The nation is something much more irrational than the State. While it is true that the State reflects the genius of the nation so that its institutions have come to be closely associated with it, yet nations have continually changed the State through which they are governed. State and Nation must be distinguished from each other. The Czarist State was radically different from the Communist State, but both have ruled the Russian nation.

But a nation can suffer from having its institutions uprooted. The violent overthrow of the State is disastrous. The French suffered in 1789 and the Russians in 1917. The suffering may have been necessary. Upheaval may have been the only way of ridding those countries of an intolerable régime. But if revolution is sometimes a lesser evil, it is nevertheless an evil because much of the virtue of life depends on accepted standards and accepted institutions: standards and institutions which are not above criticism, but which can be adapted empirically to serve new needs. Liberals believe in rapid adaptation, not iconoclasm. They welcome radical inquiry, but they see violent revolution as a sign of failure: the failure of the State and Society to meet the needs of the Nation. The nation has its own traditions and it should be used where those traditions are useful. Rivalry between nations is useful in art and in the general art of living: it is not useful in politics.

The task for Liberals is to give back power to the individual and to reform Society, the State, the Government and the Nation so that they may serve individuals. This means having not only a personal philosophy but a political philosophy, a philosophy of Government and what it should do in the twentieth century. The rival to Liberal Political Philosophy is Socialist Political Philosophy: Tories have never had any serious philosophy of their own. In the next chapter I contrast the Liberal with the Socialist outlook.

2

Liberal Political Theory

The ideal of a free or open society has been explained by such writers as Lippman and Popper. I find myself very much in agreement with them and their criticism of 'Holism', 'Historicism' and 'Hegelianism'. To me Hegel is unintelligible in theory and calamitous in practice. I go a long way with Popper in his plea for 'social engineering' and empirical intervention by the state.

Since the eighteenth century there has been a tendency for mankind to abandon the idea that they can influence their destiny. The advance of scientific understanding, Darwin, Marx and the psychological theories of the Viennese schools have been understood to strengthen the case for Determinism. This is not the place to argue again one of the oldest controversies in the world, Free Will versus Predestination. The controversy is always breaking out in different forms. In politics it takes a peculiarly perverse form. For it is the mystics who preach the inevitability of their particular creeds. It is Communism or the Third Reich which has been said to be the predestined and inevitable form of human society. The Rationalists who believe in the possibility of man controlling and improving his lot are the people who are prepared to apply scientific methods to that end.

Whatever the ultimate philosophical arguments, two lessons stand out in politics. The first is that man must act as though he could influence his destiny. Further, he at least for practical purposes can do so. The upshot of experience in my lifetime is that a great deal can be done by mankind to avoid disasters. There was nothing inevitable about the 1939 war. It is impossible to attribute it to the inevitability of the struggle between war-mongering capitalists. The capitalists, of whom Mr. Chamberlain was an

LIBERAL POLITICAL THEORY

excellent example according to the Marxian prescription, leant over backwards to avoid it. Nor is there any evidence that there were hidden economic urges towards war. As Mr. Churchill has said, the 1939 war was the unnecessary war. It was consciously willed for political motives by Hitler and his Germans. It may be that if other people had behaved differently the Germans could not have had the motives which in fact they had. But this still refers the war to human decisions and in no way absolves Hitler and his supporters from their responsibility for the ultimate deliberate decision on world war.

To take a less striking example: if we look back on the thirties we see that there was plenty of advice on how to cure unemployment; advice that would have succeeded had it been practised. The Liberal Yellow Book was there for all to read. It was the assumption that unemployment was unavoidable which proved so damaging.

The second lesson is that the approach of Holist philosophers almost always proves disastrous. Any pragmatic test shows the superiority, in terms of human happiness, of the empirical over the Holist approach to politics. By 'Holist' I mean to cover several views which usually go together. First, the view that human history is moved by vast dark forces which we can barely understand and in the main cannot alter. Secondly, the belief in the superior reality of abstractions. 'Holism' is bound up with Platonic Ideas, Oversouls, Spirits of the Age, Geniuses of the People, etc., which the Holist persuades himself have more importance and indeed more reality than individuals. Admittedly, these abstractions are often endowed with human characteristics by way of analogy only: but the pathetic fallacy is so intertwined in Holism that it is difficult to disentangle the analogies from the literal beliefs. Thirdly, Holism places intuition, emotion, unreason, above Reason which it affects to despise. It also prefers force to argument. Fourthly, it is always willing to sacrifice the present to the future. It is certain about what our children and grandchildren will want. It is confident that since it appreciates the great force of history it can show us the ideal world. To achieve Utopia it is willing to impose every sacrifice on the present generation. Indeed, it seems to take a delight in imposing these sacrifices. It often has a desire to inflict punishment on the present generation. 'Holism' —an ugly word but I do not know a better—does not necessarily embrace all these ideas; sometimes, too, as in Plato, it is developed with great subtlety. Some 'Holist' philosophers have contributed valuable stimulation to human thought, but on the whole they are the antithesis to Liberalism, and—wrong.

The most recent exponents of Holism have been the German Hegelians—Hegel himself and Marx. They have been responsible for the adoration of the Nation-State. Endowed with superhuman qualities the State has been treated as above considerations of morality. Justification for the most ruthless actions by the State has been sought in historical success. The worship of blood and iron, the German military leaders before the first war and Hitler between the wars are the legacy of Hegel. The doctrine has certainly not increased human happiness. On its own test of success Hegelianism has not been justified.

I would not accuse all Socialists of being Hegelians or Holists. But the intellectual arguments for Socialism still owe a great deal to Karl Marx, a believer in the inevitability of history, Holist in most of his tendencies and a pupil of Hegel—though to my mind a thinker in a different and far superior class.

The development of Marxist Socialism is interesting. It is not necessary to accept the Ricardo-Marx theory of economic value to be impressed by his argument that the Capitalist Society must lead to the increasing misery of the workers until they can stand it no longer. The argument is attractive. Marx's study of the conditions and tendencies of his time is impressive. But it all proved wrong. The working men have not been degraded into increasing misery, nor has the crisis of the Capitalist system arrived.

This has been very worrying to many modern Socialists for several reasons. First, because of the very strength of Marx's argument. As Socialists tend to believe in the possibility of accurate forecasting (the whole of Socialist planning depends on this) if the argument was right, how can history have gone wrong? Secondly, the failure of Marxism is a blow at the most luxuriant root of Socialism. For while the Christian Socialists can attack the Profit Motive on moral grounds, the generality of Socialists, and all those influenced by Marx, object to it at bottom because in their view it leads to greater and greater inequality and because it should not work. But what if it does not lead to greater inequality and does work better than Socialism?

The most popular way out of the distressing success of the capitalist system is to call in democracy. It is said that Marx has been falsified because the increasing political strength of the working classes has enabled them to subdue the system, and increase their share of the proceeds.

So long as this argument contents itself with pointing out what education, the growth of a more humane public opinion, the Liberal Party and very possibly the Trades Unions and Universal Suffrage have done, I am inclined to think it is partially right. The effect of giving a privileged position to the unions is difficult to assess. There are indications that recently, at any rate, their strength has been a brake on economic progress. But for my part I am convinced that the right to combine (enforced by a Liberal Government) has been, over the last hundred years, a factor in increasing the real wealth of the weekly wage-earner. For my part I am content to contrast the facts of industrial life when the unions were weak with the facts today. I know *post hoc* is not *ergo propter hoc*, but I do at least know what the unions have tried to do, and it appears to have met with success: it may be that it would have all happened without them—a Liberal Historicist would no doubt say this—but I am content to give them the benefit of a very large doubt.

Conscious effort by industrialists, working men, politicians and social reformers has improved the standard of living of the weekly wage-earners and the poor. But when it comes to claiming all the credit for political parties I am much less impressed. Mr. Asquith's Old Age Pensions and the Welfare Services started by his Government, associated with Mr. Lloyd George's budgets, and continued since, have improved the life of the poor. But the great success has been the rise in real wages and I do not think any Government can take great credit for this. What is certain is that Labour Governments have done very little indeed to raise the standard of working men. They can claim no credit at all up to 1945. Indeed, unemployment rose to its all-time peak under a Labour Government in 1930—not that I would blame them exclusively for that, but it precludes any claim on their part for improved wages or social services up to 1945. The measures introduced by the Labour Government between 1945 and 1950 flowed from all-party agreement to implement the principle of a report of a committee set up by a Coalition Government and headed by a Liberal—Lord Beveridge. This is, however, not to deny them their due both for their campaigning outside Parliament (but Liberals certainly did this as well, e.g. on Family Allowances) nor for pushing the Beveridge Report on to the Statute Book.

The facts do not therefore support the contention either that it is the pressure of parliamentary parties which has led to the rise in the real wages of the working classes or that it is the Socialist Party which has gained for the working men a greater share of the national product.

The argument that Marx has been deliberately falsified by political action has interesting repercussions whether it is right or wrong. First of all, of course, it could never have been accepted by Marx himself who believed that politics were determined by economics, not vice versa, and rejected the possibility of conscious interference with the dialectic of history. It really knocks the bottom out of Marxism for the sake of establishing the utility of a Socialist Party largely based on his teaching.

It also brings us up square against the current Socialist view of Profits and the Capitalist System. It would seem that most Socialist intellectuals still regard it as a bad system. Some think it is morally bad, others that it is inefficient. They are, however, for the most part prepared to tolerate it on certain conditions. Whether they are prepared to tolerate it indefinitely on these conditions I do not know, probably some yes, some no. One of these conditions is that its monopolistic characteristics be accepted as inevitable and on the whole desirable, but made subject to stringent parliamentary control. But it must be said that there is a direct clash of opinion here. Some Socialists have taken a leading part in trying to curb restrictive trade practices: in any Finance Bill a plea for small businesses can be heard going up in the most chivalrous accents from the Labour benches in the Commons. At the same time the counter argument can be strongly heard to the effect that the 'oligopolist' tendency is inevitable in industry, modern research and techniques require big units, expansion must come out of reserves, and indeed the argument that large units are more stable, more easily controlled and easier to nationalize. The most important condition, however, is that industry should be under even more stringent public control.

I call this the wild bull attitude towards capitalism. It seems to imply that capitalism is a fundamentally dangerous beast but that it can be put to tolerably good use in the most severe straitjacket.

It is a wholly mistaken view. If the choice were really between running a free enterprise system which was fundamentally unsound at the cost of continual State interference (which must mean the surrender of greater and greater powers to the State) and having a Marxist revolution followed by a new and better economic system under which the State could wither away, I would choose the latter.

I find the present neo-Marxist position quite unsupportable—far more unsupportable than the genuine Marxist position which at least pronounced that the State would disappear.

But why do Socialists cling to the view that the capitalist system, with its profits, is unsound? Whatever evils existed under it in nineteenth-century England, they were not nearly so bad as the evils under Socialism in twentieth-century Russia. Undiluted capitalism was humane compared with undiluted Socialism. It is not capitalism that is the mad bull, but Socialism. Society under capitalism performed remarkable feats in the increase of wealth and has reformed many of the admitted evils which existed. Perhaps Socialism also will reform, but our experience of it is that it leads to oppression within a country and war between countries.

The other argument sometimes advanced for the failure of Marxist prophecies is that the doom of capitalism has been postponed by its ability to exploit its colonies. I do not think this need be dealt with at any length. There is no historical evidence for it. Indeed, it could probably never have been invented were it not necessary to find some excuse for the master. But what is particularly interesting is that as the colonies emerge from under the heel of the capitalist, where according to many Socialists they have writhed so long, they show all the same desires, conflicts, and indeed imperialist leanings that were condemned in their old colonial powers.

Cut away from its Marxist root it is difficult to find the philosophical basis of modern Socialism. There are still some moral Socialists who believe, in varying degrees of course, that profits, competition, wealth are all wrong. They are perhaps the weakest, but also the most lively element in the party today. They are difficult to explain or criticize because their view is, perhaps necessarily, incoherent. They engage in many protests, for which I do not blame them, but some of their complaints seem difficult to relate to their main views. For instance, they are nearly always on the side of the strikers in a wage dispute in a nationalized industry. If the strikers were oppressed by bosses, one could understand this. But the bosses are gone. The employers are the public, according to Socialist theory. There are no profiteers or profits. Such Socialists seem to draw a moral distinction, which I find difficult to follow, between the efforts of a wage earner to increase his wages, of which they approve, and the efforts of a shopkeeper to increase his takings. Again, this type of Socialist is nearly always willing to see the power of the State increased though they are often at the same time deeply concerned about infringements of liberty. But it is often the State that infringes liberty: and the other powerful organizations which occasionally threaten our liberty frequently work in conjunction with the State and are usually acceptable to Socialists, and it is the Nation State, too, which they want strengthened, which has been most belligerent. The profit motive and competition have their bad side. I do not deny that they can be abused. However, the pure *laissez-faire* system never did or could exist. The power of money has been exaggerated. It can only exist, it can only be exercised if the civil authorities allow it. But certainly great unearned inequality of wealth, or the power to use wealth as a force, like a bully, are bad for a society and the State should intervene to curb them. On a higher plane, too, there is something most attractive about a world in which the motive is love and devotion to an ideal. It would not lead to equality but it is instantly seen as good —better than personal gain.

Nevertheless, this is not to say that the profit motive is necessarily any worse than the wage motive or the pension motive or the National Assistance motive. Profits have a bad name because they still have the smell of 'unearned' usury about them or the taint of exploitation. But the man who saves and invests and waits and risks performs a service. Nor is a 'profit' on a transaction necessarily any more exploitation than a demand for National Assistance. Either may be justified or unjustified. But to be eager to get as much as you can out of your fellow-citizens through higher wages, pensions, National Assistance or tax evasion for that matter, can be due to just the same motives, and subject to just the same moral objections, as profits.

Holding as I do that Governments are better employed promoting happiness than in teaching morality, though I trust they will safeguard the circumstances in which morality can flourish, I am prepared to use the profit motive, ambition, selfishness, call it what you will. It is often used to improve the condition not only of the individual concerned, but of his family, friends and neighbours: it has been the inspiration of many acts which have benefited mankind. Nor is it different in kind from some of the springs which drive men to seek positions of power and responsibility to which even the most self-effacing moral Socialists often aspire.

But, of course, the great middle run of Socialists are neither logical Marxists nor consistent moralists. If they do not approve of the present profit system they are very willing to cash in on it. They have lost faith in the motive of public service and in Socialism. And now it is the policy of their party as a whole that the State should hold equity shares and take profits from industry. Indeed, some go so far as to admit that there is nothing wrong in profit. Once they reach this position it seems that they should accept the success and serviceability of the system. It then is necessary to see how the system can be worked to the best advantage, its virtues admitted, encouraged and exploited. This is a long way from the mad bull view of capitalism. It is the end of the road for Socialism and the beginning of Liberalism.

When it comes to a discussion about how the system is to be worked and who is to benefit, then we are far away from Socialism either Marxist or moral. This is a topic on which Liberals have a great deal to say.

But in their obsession with the economic and materialist side of Socialism, the Labour Party has often been blind to its political implications. The concentration of power in the State whether by direct nationalization or by cashing in on the free enterprise system is both inefficient and dangerous. But the belief that the State can plan on a vast scale is even more dangerous. One asset of the free enterprise system is that decisions are taken by a great number of people. That alone is a safeguard against one overriding disastrous error. These decisions too are for the most part taken by people who have some knowledge in the field in which the decision lies. They are taken by people who are interested and who are responsible. They will feel the results.

The belief in the possibility of overall planning rises from a 'Holist' outlook. It overlooks the multifarious needs of a community, it forgets that abstractions seem manageable because they are unreal. The individual is far more difficult to deal with, unless he is treated not as an individual but as a 'hand', a statistic, a member of a class.

The approach of those who regard politics as 'social engineering' is far more sympathetic to Liberals. Indeed, Liberals are social engineers—but engineers with firm views about the engine they are trying to keep in running order. For 'Social Engineering' political theorists the removal of faults in the political system is the first task. They are empiricists trying to bring reason to bear on the problems which are felt by individuals. They are at once more humble and more optimistic than Socialists. They are more humble because they are prepared to learn from experience that what one man or one generation may think good for his fellows or for their successors is by no means always appreciated by those who are supposed to reap the benefits. They are optimistic because they have faith in human action based on examination of the facts available to them. They are not so defeatist as to give up in the face of situations which do not conform to their preconceived theories.

But empiricism, though essential, is not enough. We are reaching a stage when the 'given' in the world of mankind is becoming much smaller. The length of a man's life is increased. Greater control could be exercised over man's numbers. Human beings may soon be bred with certain characteristics, determined by their parents. We may learn to control the weather. All this raises again questions about the ultimate ends of human life. At first sight the whole advance of science seems to be towards a kingdom of robots from which humanity is being expelled. But this is a superficial view. The greater the power we have, or seem to have, over our destinies, the more important it is that we should exercise that power rightly. The bigger, too, is the scope for error.

The 'social engineers', those of them who are pure empiricists, are right to emphasize that the way of advance lies in applied reason. They are right to stress the need to remove faults rather than to build Utopias. But the existing situation needs drastic change. Change which is so drastic that doctoring here and there is not sufficient: it may indeed make matters worse. In fact, a great many of our troubles have come from tinkering about with things which should have been cut away. There are many examples. We tax the cinema industry and then when it gets into difficulties we give some sections of it a subsidy, instead of abolishing the tax. Our tax regulations themselves have become a complete cat's cradle by a species of piecemeal engineering. The advocates of social engineering (of which, broadly speaking, I count myself one) would say that these may be examples of bad engineering. They do not, so the engineers would claim, invalidate the argument for good engineering any more than the possibility of a garage ruining a car when trying to repair it should lead people to scrap a car as soon as it breaks down.

But the mechanical simile is enlightening. For you must know what you want the machinery to do. So 'political social engineering' depends upon some decisions about what is ultimately desirable in a society or in a Government. I do not mean that it is necessary to have a complete platonic view of the ideal society before you can remove blemishes from, or graft improvements onto, our institutions. But it is nevertheless necessary to have some general, even if alterable, idea of what you want from the instrument in hand before it is possible to carry out any repairs either to a car or a system of government.

Further, and this is the compelling reason for having some skeleton of the ideal society in mind—no politician can summon up the energy to do the necessary repairs without some vision of a more happy world to be created. Over the years an immense advantage of the Hegelians, the Platonists, the worshippers of the State, has been that they have been given the courage to attack all difficulties by the ideal they have in view. That is what has made them progressive: they have had the will to move the heavy inertia which lies all around us. Unless the social engineers can counter this with a vision of their own, very little social engineering may in fact be done. Professor Popper and others seem to me to underestimate the friction which slows up any conscious political change. My experience in politics is that the Conservative empiricists in Great Britain, though they may groan and grumble about the state of the world, will not summon up the determination to change it. They will wait until some crisis, or the gradual build-up of events, forces them to move. They will then move as little as possible. Conservatism has its own brand of inevitability. For them it is not the inevitability of change. It is the inevitability of the present position. They conserve no set of principles, but rather the state of affairs they inherit.

The politician who believes that improvements can be effected by the conscious application of reason to the world around us needs some stimulus other than simple intellectual conviction that an improvement can be made. He needs the inspiration of a vision. Not, I hope, a vision of a heaven on earth tended by a Government no matter how beneficent, but a vision of the conditions in which men can have a choice, responsibility and scope. No policy is of any value without the will to push it through. The will is unlikely to be steeled by intellectual argument alone.

A further danger in the piecemeal approach of social engineering is that the engineer will actually do the wrong thing. This is largely due to inadequate political institutions. In the chapter on political reform I suggest that the pressures behind political action have been insufficiently explored. I argue for a reform of existing institutions and methods and the development of new types of democracy. If social engineering is to be successful it is essential that these reforms should be made. But some vision of where we are going, some principles of action are also essential. Anyone who watches the House of Commons at work can see the mistakes social engineers make when these principles are abandoned. If there is unemployment in industry 'A' why not give it a small subsidy: it seems a useful piece of repair work, well justified by its results in human happiness. Then if industry 'A', why not industry 'B', and before you know where you are, you are through the alphabet of industries. I am sufficient of a social engineer to be on my guard against the opposite fallacy of saying that we can never help industry 'A' without letting down all the barriers of reason and helping every industry in any difficulty, imaginary or real. Human reason ought to be able to differentiate. It should not be necessary to have, in day-to-day political matters, unthinking adherence to rules of thumb. But a practical study of how democracies work shows what muddles the empirical approach can in fact create. The answer lies partly in resisting the pressures which build up—but partly also in having some fairly clear idea of the society we want to create. Without such an idea it is all too likely that social engineering will mean taking the sound tyres off the motor-car and selling them for short-term gain rather than mending the puncture.

3

Liberal Political Reforms

The root of our troubles is political. One of the greatest contributions which Liberals can offer today is a contribution towards political reform.

Democracy can be justified on three grounds. The first is that man is a political animal. The exercise of his political talents is a good thing. Even if a perfect tyranny could be guaranteed—perfect in the strict sense that it answered all its subjects' needs except their need for political expression—it would be bad. The second two are based more on experience, or expediency, than on ultimate judgments about the ethical value of Democracy. Under this head it is said that Democracy is the least bad form of government, counting heads is more civilized than breaking them, and taking decade with decade Democracy works best; or that even if Democracy may not be the least bad form of government, it is the most acceptable.

I am prepared to accept all these arguments, for I do not believe that they are incompatible: nor do I see much purpose in a very theoretical discussion about some totally new way of running our affairs. A preponderantly popular form of government has been an integral part of the British tradition for some time and that is yet another argument in its favour. I accept, too, that Democracy means that ultimate power rests with the people. But having reached these two conclusions almost everything else remains open.

Democratic government can take hundreds of different forms and raises many problems, some of which, such as the representation of minorities, have exercised political philosophers for a long time. But all forms depend on some system of popular vote. And

LIBERAL POLITICAL REFORMS

it is usually also assumed that while you may want to devise more delicate systems of voting to give proper weight to different opinions you cannot go behind the vote. John Stuart Mill, indeed, argued for a minimum standard of education—that citizens should be required to learn how to read and write before they were entrusted with the vote. He also suggested that a most important function of democracy was to exercise people in politics. But neither he nor anyone since has given much systematic thought to the pressures which are built up before anyone casts a vote.

Most people vote, I suppose, from self-interest: some from enlightened self-interest, looking to what they believe will be good for the country as a whole: most from rather short-sighted self-interest. Their view of where their interest lies may depend very much, however, on the political system. Still more is the active politician's view coloured by those interests which can exert the heaviest pressure on him.

Let us take two very different examples of political pressure groups: the Trade Unions and the Lord's Day Observance Society. The former are frankly and properly concerned with getting as much as they can for their members. This does not mean that they are out to destroy the rest of the community; on the contrary, in proportion as their self-interest is enlightened, they will moderate their pressure for immediate and often illusory advantage for the sake of the more solid but remote benefit of a better society. But no one who studies the language of Trade Union leaders or thinks over the recent history of their pressure for more power and higher wages, pressure sometimes pushed to the point of strikes most harmful to the public at large, can doubt that short-sighted self-interest sometimes wins the day. Trade Union leaders sometimes treat employers and indeed the community at large as people out of whom their members must screw or bully advantages. They have used the weapon of coercion to further their own power against other unions, to press individuals into their fold and to win benefits for their members at the expense of their neighbours. At the same time they occupy a most important position politically. They take a levy from their members out of which they must explicitly contract if they do not want to subscribe to the Labour Party. They contribute to the support of about 100 Members of Parliament. They claim that the seats for which these members sit should be treated more or less as pocket boroughs.

All this is done by a body which has special privileges in the State. Liberals are in favour of Trade Unions: they think it is a good thing that they are industrially strong. But the political results of their privileged position are bad. Just as we would not now tolerate, say, landlords or churches which claimed special privileges under the law, sought to make it a condition of work that people should subscribe to their funds, used their funds for political purposes unless a man explicitly contracted out and supported a solid body of representatives in Parliament, so we should revolt against these practices by the unions. But I am not suggesting that abuses are confined to Trade Unions alone.

Take next the Lord's Day Observance Society; its views are held by at most a fifth of the English people and put into practice by even fewer. If any sort of referendum were to be taken on Sunday Observance, a large majority would be found against the existing laws. But Parliament has never been compelled to change them.

The influence on politics of the Trade Unions and of the Lord's Day Observance Society is not simply due to the fanaticism of their members or the efficiency of their organizers. The extent of their power is partly due to our political system which has done nothing to redress the unbalance between what the majority of people want without zeal and a few have a fierce or vested interest in procuring.

The single member constituency in which it is the simple majority that counts does not, as is sometimes said, mean that what most people broadly want prevails against the minorities. On the contrary, a united block within the dominant party can exercise pressure out of all proportion to its size. It can do this either within a party or by whipping up its members to make their wishes known to the M.P. direct. If they can give the appearance of either being marginally decisive or of being the tail which can wag a too indolent dog, they have every chance of success. Further the present system of election maximizes local and immediate issues and minimizes long-term questions affecting the community as a whole. The single member constituency is often dominated by one

LIBERAL POLITICAL REFORMS

industry. Its representative may become a prisoner of that industry. If Lancashire, including Manchester and Liverpool, were one constituency then it would be an area of very varied interests. Cotton would be important but not supreme. Members would perforce have to take a broader view and embrace many more varieties of constituent than they do at present.

The single member straight majority system also encourages indifference to politics. For in many constituencies the result is always a foregone conclusion. It is not only the minority who are frustrated by this—but the majority.

On two major tests, therefore, our present system fails. Instead of lending weight to the general long-term considerations it magnifies selfish and narrow political pressures. Instead of making as many votes count as possible it makes as few.

We should not be content to go on with a system which is in no sense democratic, which distorts the will of the people and virtually disfranchises great sections of them and which also builds up the wrong pressures. I doubt if the situation is appreciated. Its faults are not only that it disfranchises Liberals. It is plain, downright inefficient. The party with the most votes need not win the election even if only two parties ran, and the odds are much against it doing so if more than two are in the field.

What makes the system ultimately intolerable is its secondary and less obvious effects today. Combined with the growth of 'closed-shop' politics, particularly on the air, combined with the strength of the Conservative and Labour machines and their dependence on monetary levies from either side of industry, it has become an instrument of reaction. The present electoral system fosters the present party system. But it is also greatly aggravated by it. Its insistence on discipline in the House of Commons, the carrying of national party politics into every corner of government and the decline in the number of people who are in a position to show any independence in politics have all accentuated the general evil.

It is a further count against the present electoral system that it denies any place to the independent. The members who were throughout their career wholly independent of party have not, in fact, played a very distinguished part in our political history. The

LIBERAL POLITICAL REFORMS

university seats provided one or two notable members. But on the whole it is depressing how often they returned party hacks and how little impression even their more enterprising choices have left. However, independence is worth cherishing if it only gave us Miss Rathbone and A. P. Herbert. And we do not know how many more bright spirits have been lost to politics through the dead hand of the system even when there was through the university plural vote a small chink in the iron curtain. But it is not only the individual independent who is being squeezed out. It is the independent-minded Tory or Socialist. And it is the group of dissidents. Today, if they were starting, it is doubtful how far Lord Shaftesbury or Sir Winston Churchill would get. And as for Fourth Parties and Clydeside Groups they would be quickly and effectively driven to the Chiltern Hundreds.

The practical steps which are needed to remove the distortions of our present political system and restore it as a vehicle for a Liberal Society can now be summarized.

The total dependence on the single-member simple majority constituency must go. The single transferable vote in multi-member constituencies seems, for most places, to be the answer. There are some areas, such as the North of Scotland, where the population is so sparse that the creation of multi-member constituencies would meet intolerable difficulties. In the extreme north to create a constituency which in comparison with southern figures would come near justifying three members would mean taking in an area stretching from Inverness to Flugga Light (about as far as from London to the Scottish border). Further, so long as the system was altered to meet the evils flowing from the present situation Liberals are not doctrinaire about how it should be done. There is much to be said for the Alternative Vote and for various other systems. Such a constitutional matter should, if possible, be discussed with other parties and concessions made to their views even if ultimately Liberals must maintain their view that the present system must be reformed.

We should forbid the sponsoring of Members of Parliament by outside bodies: if such bodies want to contribute to party funds their members must contract 'in' not 'out'—and this should apply to the companies which support the Conservative Party through the 'Aims of Industry' as much as to the Trade Unions. Shareholders who want their companies to take part in politics should say so explicitly.

Liberals should also be concerned with the way in which issues are considered and opinion formed. On most political questions to reach an opinion by reasoning is laborious. The number of questions raised and their importance grows all the time. Yet we are not teaching people to think logically. The weighing of evidence and the following through of an argument are largely neglected in our schools. Parliamentary democracy and certainly Liberal parliamentary democracy depends a great deal on the habit of splitting 'what I want' from 'what is true' or 'what is good for the community as a whole'. It depends upon the reference of decisions to certain standards and not merely to emotion. But we take little trouble to see that this difficult process is practised. I doubt very much, for instance, if many people in the country grasp even the elementary considerations for or against the Sterling Area.

But there is the further question of political information. The main medium is now television. Yet we have allowed the Conservative and Labour Parties to clamp rigid blinkers on to the viewer so that he is rationed to carefully prescribed doses of political information heavily weighted in their favour.

I come now to Parliament itself.

Parliament is groping towards realization of its inadequacies. It is more and more common for Royal Commissions or Select Committees to be appointed even on subjects about which politicians might be supposed to have strong and informed opinions. We now have the curious phenomenon of the Cohen Committee which merits a little consideration all to itself. Nowadays we are supposed to have a managed currency. It is admitted by all parties that the regulation of currency and credit is a function of the Government. Liberals at least would say that the maintenance of reasonably stable money is a prime responsibility of the Government. Yet since the war we have suffered from continual inflation and monetary disturbances as serious as any in recent history. Nor is there really much mystery about what has been happening. If a Government had wanted to stop inflation they could have done so. But the political pressures to continue it were too strong. The forces of the citizens organized as producers, as addicts of the pound in the purse, were much stronger than the vague realization that inflation in the long run, if too luxurious, will rot us all. The executive, in the shape of Tory Governments over the last eight years, has proved unable to do what it said it ought to do. So on this supremely political question, in an age when more and more questions are being brought into politics, the Government runs for cover behind an ostensibly non-political body. It tacitly admits that the wrong political forces are too strong for it. It also tacitly admits the need for public education and for some body, some group of persons, with a prestige greater than its own. Can anyone imagine Mr. Gladstone setting up a Cohen Committee on Ireland or Mr. Churchill on Foreign Affairs in the 30's?

Then there are the effects of the methods by which we carry on parliamentary government. The budget is now the instrument which bears most directly on people. Not only does it settle their taxes, it is the weapon by which the Government seeks to impinge on the general economic tempo of the country. Yet it remains in form very much as it was a hundred years ago. It is annual, but economic changes do not conform to the twelve-month cycle. Twelve months is even too short a time for much of the government accountancy: the threat of annual changes in the Purchase Tax, for instance, is most damaging. The annual budget rests on the fiction of Treasury control, which in today's world of inflation and vast government expenditure does not exist. It is composed in a way which makes it as difficult as possible for people to know what they are paying and for what. It is a muddle: a muddle between capital and current account: between a balance sheet and an expression of economic intention: between a general stock-taking and the examination of various taxes, their effects and their practicability.

It is not only for efficiency that parliamentary methods of handling finance must be altered: it is for the sake of public understanding of what is going on. If they were asked, I doubt if many people in the country would be in favour of the subsidy to airports. Indeed, if the loss of money from their pockets could be directly related to the purposes for which it is spent I doubt if a majority of people would support anything like the present rate of subsidies.

LIBERAL POLITICAL REFORMS

The pressures which are built up in British politics which tend to favour sectional interests also favour State action. The habit of demanding that the State shall run everything has grown astonishingly, and though in some ways it is very obvious the general tendency has been strangely unchecked. When we hear that the State should subsidize sport, see that shoppers are only offered quality goods, regulate prices of all sorts and settle industrial disputes, it no longer shocks us. After all, it already regulates a great many sales, says when shops may open or shut, and has an elaborate machinery for industrial conciliation. It is manifestly very much easier to get the State to dole out more money and assume new duties than it is to persuade it to abandon or retrench its activities. And the reason for this is not entirely a change in what people want—though that is a very powerful factor. It is partly that in the course of two wars and a century of collectivist thinking we have never paused to think out the proper sphere of State activity nor the causes of this rush of electors to turn themselves into a proletariat. I suspect that the slither towards dependence on the State is inherent in all democracies unless deliberate steps are taken to counter it. It led to *panem et circenses* in Rome: it is a constant fear of political writers.

I see no reason, however, why steps should not be taken to counteract it.

In countries like America which have a written constitution the power of the Government is circumscribed. That would be contrary to our traditions; indeed, it would be difficult to clamp a written limitation onto the tradition of the British Parliament. But there are other ways in which the Government could be compressed within a more legitimate field. A second chamber, elected but less acutely dependent than is the House of Commons on the favours of the electorate, would achieve something. The weakening of the party system which might be achieved when the single member constituency is altered would give a better platform to voices and votes not wholly at the beck and call of producer interests. These I consider later. But easiest of all would be some change which would force members to give more thought to the implications of what Parliament does and to take more direct responsibility for what they propose.

LIBERAL POLITICAL REFORMS

Parliament is the traditional watch-dog which should be ready to drive an encroaching executive from our liberties. Most particularly it used to be the check on public spending. For the imposition of taxes has always been a handy weapon for tyranny. Now all this has been turned upside down. It is true that a minority of members are roused by infringements of personal liberty. But so far from restraining the executive or curbing public expenditure, Socialist members are committed to higher and higher taxation and indefinitely increasing public control, while Conservatives have achieved no reduction in total government expenditure over the last eight years nor do individual members of any party, with a few honourable exceptions, show much reluctance in pressing particular items of expenditure on the Government.

It is not only over finance that Parliament is in need of reform. Nor is it only by our methods of taxation that we encourage the wrong tendencies. It is high time that we had a fresh look at the purpose of Parliament today and its relationship with Government and with the people. Parliament is the legislature, or at least in name the main part of it.

But except for a few private members' bills of small importance, the House of Commons does not initiate legislation. Nor are its members equipped to consider legislation. They have no offices as have American Senators. The whole organization of Parliament is bent towards legislation by the executive, amended only to a minor extent by either House. The rigid party system under which we operate turns the so-called legislature into at best a convenient place for revision by the Government.

The main function of the Houses of Parliament is now to provide a clearing house for ideas and grievances of the electors and a forum for policy announcements by the Government. Matters of public interest can be aired, the Government is subjected to constant criticism in public, there is public debate. It is of crucial importance to this process incidentally that the Ministers should be Members of Parliament and so while we never elect our Government by direct popular vote we do at a general election provide through the House of Commons the main body from which Ministers must be chosen.

But here, too, its authority and usefulness are being eroded. And here, too, the present party alignment is much to blame. For several of the most important issues do not lie between the parties but divide both the Socialist and Conservative parties. Neither, therefore, wants such issues publicly debated in the Commons. Indeed, more and more are such decisions as can be said to be taken at all by M.P.s taken in the secrecy of the Committee Room; a proceeding which by a stretch of words can be described as democratic but which is essentially a Communist or one-party type of democracy. There is also the custom by which pressure groups by-pass Parliament and go straight to the Minister. Nor is the approach always made by the pressure group. The conversations which Ministers and their officials seek with interested parties such as local authorities, Trade Unions or Employers' Federations are often far more decisive than what is said in Parliament.

Another restriction on the usefulness of Parliament as a forum arises from the curious conventions about what should be discussed. The estimates are never properly discussed. The nationalized industries are most inadequately discussed and their relations with the responsible ministries or the work of those ministries themselves are hardly mentioned. It is considered irresponsible to debate many questions which disturb the public, and even more irresponsible to put forward solutions which are controversial. For instance, it would be difficult to debate restrictive practices by Trade Unions in a reasonable atmosphere. And how many times have we not heard that hoary, old and nonsensical appeal not to divide the House, because such a division will exacerbate someone's feelings. Not even Chesterton's satire has killed that well-worn absurdity of Ministers winding-up a debate. The situation is worst in Foreign Affairs and Defence.

Recently the Tory Party have been exceedingly unwilling to discuss their foreign policy: partly from the lack of any viable policy in the Middle East, partly from an exaggerated fear of indiscretion. The Labour Party have had no particular relish for a foreign policy debate, because they, too, have not been certain where they stand on several issues; and on defence, which is apt to become involved in foreign affairs, they are deeply divided. The Liberals, as a party, have no parliamentary time. The result has been that from February to July 1958, the House of Commons virtually had no discussion on foreign affairs. The Government were not compelled to expose their Middle East policy to scrutiny. The people were not informed on the issues. Even when troops were pouring into Cyprus and every newspaper carried forebodings of an American landing in Lebanon no debate took place. The Secretary of State for Commonwealth relations solemnly rebuked a peer who dared to hint that the troops were being sent to Cyprus with a possible view to action elsewhere. Within a month some of these troops were in Jordan. Few people outside the Liberal Party, however, felt that the Government or Parliament had been in any way lax in not giving a rather truer picture of the situation and of the Government's intentions.

There is a similar conspiracy of silence and indeed misrepresentation on defence, on the nationalized industries and other subjects. No sensible person can really be sure that by profitable trading in the market without subsidies, cheap loans or other assistance, the transport industry is going to break even, taking one year with another. The deficits mount. The Government neither tells us how they are to be paid off nor takes steps adequate to stop the rot.

What ought to happen is that the Government should explain —or be forced to explain—to Parliament what its policies are, and should then be hammered when it flagrantly abandons them. The main lesson of Suez, for instance, is that the Conservatives went on preaching doctrines at and between elections in which they had no faith, and which they abandoned when the test came. They put no trust in the Tripartite Declaration or the United Nations when war broke out. Now whether their ultimate action was right or wrong, they should have been called to account for proclaiming a policy in which they did not believe, or which was found wanting. If Parliament is incapable of extracting information from the Government on vital matters, if the Government brazenly go back on their electoral statements of policy without penalty, the public will become cynical. The House of Commons will have manifestly failed in another of its chief tasks. But now when party unity, though difficult to achieve, is all-important for the party leaders, and when some seventy members of the governing party are in the executive and more either are, or hope to be, recipients of its patronage, the House finds it difficult to fulfil this function of extracting policy intentions from the executive.

So we have too little instruction by debate and it is often too late. Coupled with the muzzle enforced by politicians on T.V. this, again, means that the pressures which ought to drive democracy to attend to the wider and important issues are lessened.

We need a stronger Second Chamber with no hereditary element and free from dependence on the executive. It might be elected by very large regional constituencies, or indirectly by regional bodies.

We need new parliamentary procedures including the establishment of standing committees and more time for private members' motions. But, above all, we need new parliamentary conventions. I am not sure that the suggestion made by Mr. Christopher Hollis that the power of the whips is a breach of privilege has not more in it than an agreeable joke. There is a need for cohesion in a party, but the present predominance of the party is a danger. We need the custom established and enforced that the Government explains its policies to Parliament and as a general rule sticks to them. Standing Committees following the trend in such subjects as the foreign and colonial affairs and the nationalized industries would be most valuable. We want a thorough overhaul of Parliament's purposes and responsibilities. We want much of the work which must be done by public bodies devolved to Scotland and Wales, new regional local authorites and functional bodies.

As well as some attention to the pressures behind the vote and indeed behind political action of all kinds, as well as reforming Parliament itself, we must also look at the other channels through which democracy must work.

Sovereignty in this country at present is believed to lie with the people at large. But in fact the people have already pooled much of their sovereignty with America and their Western Allies. They ought, in the eyes of Liberals, to pool more. This subject is dealt with in succeeding chapters. Here I deal only with our internal situation and the organs through which we should exercise democratic control over the operations which concern us alone. Of course, the distinction is not an absolute one: indeed, Liberals must insist that the old breakdown of our business into home and foreign affairs has become unreal.

LIBERAL POLITICAL REFORMS

The exercise of sovereignty by a democracy has of late received almost as little attention as the pressures which move the democratic will. The purpose of debate as a prelude to decision is to let everyone have his say; human nature demands some freedom of expression and mere letting off steam is not to be despised: but it is also to assist in coming to a right decision. Half a dozen minds brought to bear on some subjects are at least less likely to make gross errors, if not to achieve brilliant conclusions, than one mind. But the purpose of debate has only to be stated in this way, as a means of finding truth and getting it acted upon, to rouse at once misgivings and qualifications. As for finding truth, it all depends upon the question, who debates it, how well it is posed and how judiciously the debate is conducted. As for getting results, it must be stressed that all parliamentary debate is aimed at moving someone to do or stop doing something, or to changing the conditions under which action is or is not to be taken; so the form of the debate is of the greatest importance.

The essence of democracy is neither the vote, nor the debate, it is the creation of a field for every individual in which he makes his own decisions and the creation of a system in which he can properly form his own opinions and if he wishes make them heard. If the educational system does not train him to form opinion in the right way, if the system by which information is disseminated is faulty, then democracy is imperfect. But, further, democracy has to sort out who is to give his opinion and how. There is no reason why every question in a democracy should be settled by vote or by political argument.

The test should be that those matters which affect us all and which are specifically of British concern should be decided by Parliament. They should be so decided in the sense that general rules should be laid down for them. The amendment of the general law is a proper subject for Parliament. So is the general economic framework. It is to my mind wrong for Parliament to intervene to say that petrol stations may not be built but village halls may be, unless it can be shown that petrol stations fall under some generally acceptable definition of what is bad for the community while village halls are part of a class of operations conferring a general benefit. Whether either building falls in the class stated is a matter for discussion. That is what the discussion should be about. Further, it is not the business of Parliament to make men good, but to create conditions in which they may make themselves good. (See p. 14) I doubt if it is even its business to make men happy, though I think that much more arguable. The Festival of Britain was not one of the greatest triumphs of British democracy, but the removal of poverty, ignorance, and incitements to vice has been. Our Parliament is much better, when legislating, at the negative than the positive. The positive power of Parliament lies in its capacity as a forum. It is in explaining the issues and rousing people to think and speak and act for themselves in the Gladstonian manner that Parliament is positively valuable.

Parliament, apart from presiding over the general framework within which we live our lives, can legitimately undertake specific tasks on two grounds: first that the tasks are communal, or secondly, that they are so important, although neglected, that on empirical grounds it is convenient to place them under Parliament. But before Parliament undertakes these tasks the justification must be carefully weighed. Parliament should also look to a suitable occasion to shed its work to a more suitable tool. Many tasks are put upon a Government in war. Many of these remain with Parliament in the peace. They should not. The development of nuclear power is an example of something which Parliament may properly take under its wing for a period. Indefinite, in all senses, responsibility for the aircraft industry is something it should not.

Finally, there are some things concerning the prestige, the spirit of the nation as a whole, with which Parliament, its supreme expression, must deal. It must deal with Commonwealth matters, with matters concerning the Crown, the Armed Forces, and, in so far as one can speak of them, Foreign Affairs.

It will be seen how few are the subjects which are really suitable for parliamentary debate and control. Let me at once say that this does not mean that all the others are not matters of public concern. You could 'nationalize' (if that word now has any precise meaning) industries without making them subject, except in the most remote form, to parliamentary control.

There are two great ranges of subjects which are true matters for public control or public initiative outside Parliament. There are local matters. There are practical matters. Local matters have got into a hopeless muddle. The basis of local authority was the difficulty of communication and the peculiarities of local feeling. Both have changed out of all recognition, but we pretend the situation has remained the same. There would be little physical difficulty now in administering the whole of Scotland from Edinburgh, or the whole of England from London. Even the County Councils were started before the motor car. As for the Parish and Borough Councils, they belong to the Horse Age in its bridle-path and saddle era.

In accordance with the niggling, time-wasting and futile method by which every change is put off as long as possible and the minimum is done, the recent changes in Local Government have not touched the root of the matter. They again, according to modern custom, merely complicated the over-complicated superstructure which rests upon totally inadequate foundations.

Local Government at present suffers from several major faults. Its finances are rickety. It depends for income on rates and central grants. The first are an arbitrary and out-of-date system of finance: the widest source of wealth in many industrial areas is still partly derated and while rates fall heavily on some who use local government services others virtually escape. The central grants, though essential, make a mockery of local government responsibility. Over 80 per cent of the finances of some counties are contributed by the central government. The boundaries of local authorities need revision: Sutherland, with some 13,000 people, is a county nominally of equal standing with Lanarkshire or Lancashire. The functions of local authorities also need to be looked at again.

If local government is to be revived it must be by creating local government areas which have some relevance to the matters which their councils are supposed to consider. The pathetic fallacy of attributing political needs to places has been pointed out long ago.[1] There is little loyalty to counties except perhaps in sporting or hunting circles: these can keep their allegiances: for old time's sake, too, let even the most unviable of counties remain on the map. But in our nostalgia do not forget entirely that local government, even more than central government, was made for man and

[1] J. S. Mill, *Representative Government*.

not man for local government. Larger units of government are needed between Parliament and 'the third tier'. This will mean a redrawing of areas, but what may be lost by the removal of political power from, say, the citizens of Rutland, may be gained by the recreation of the old provinces possessed of the means and the will to be real local governments and not merely shadows of London.

A Liberal regards local government as the getting together of people who have a common interest limited roughly to themselves. By this criterion education, major (and not merely trunk) roads and the wider aspects of Town and Country Planning do not fit within the boundaries of most local authorities. There is some common interest in many counties. But there is no need to pretend that the rigid pattern of local government, which has in reality broken down, still exists. Counties are being grouped for some services, e.g. Police. But the changes are haphazard. We need at least in some parts of the country a confederation of counties or a regional body which among other things should deal with education. It should also for many purposes be the Town and Country Planning Authority. I should hope that the regional bodies, though responsible either directly or indirectly to the people at large, would develop a rather more enterprising and less administrative outlook than the present local councils.

In Scotland and Wales, above the regional bodies, there should be separate Parliaments. These would take their place in a genuine pluralist society. Much of the opposition to Scottish and Welsh devolution comes from imagining small Westminsters in Edinburgh and Cardiff. To my mind the case for devolution loses much of its force if it is cut out from a general scheme of political reform and considered merely as a fifth wheel on the already over-cumbersome Westminster coach. But if Westminster is shorn of many of its functions and reconstructed as a legislature and forum for discussion designed to influence and control the executive on major matters, then there can be some genuine place for national parliaments giving expression to the effective and beneficial side of nationalism. I visualize a Parliament in Edinburgh, for instance, dealing, among other subjects, with subjects like Town and Country planning and the Arts. I see it, too, co-ordinating and financing the development functions. It should be coupled with a reform of the tax system to incorporate a tax on Land Values. It must be firmly insisted that these improvements are part of the Liberal demand for less government. Through such improvements a Government better attuned to its proper functions will also be made more effective.

Some of the vitality which we should all like to see at the grass roots, at the street and district and parish level, can surely thrust forward again on practical matters. The Quakers and Bosanquet have both told us of the 'spirit of the meeting' which can be evoked by discussion in a small gathering. As in all kinds of debate, to achieve this spirit needs attention to stage-management. It needs also a meeting where the protagonists have a good deal in common, are unbigoted and instructed in the matter in hand. The hazards are known to anyone who has ever sat on a committee. Instead of the members of the meeting recognizing freely that something has emerged better than any one single suggestion put forward, they leave bewildered, having been browbeaten or tricked by some astute committee-man into a travesty of what they intended. That is one danger. The other is the totally silent committee which, for the sake of each other's feelings, leaves equally frustrated with an equally foolish decision to its discredit. But given an appropriate subject and a good chairman with a practical matter to decide, on which opinions are not too diverse or doctrinaire, a small committee can be an excellent body. The Health Service, Education, the administration of a port, these are unsuitable subjects for mass democratic action, but one quite suitable for democratic action by those who are interested and knowledgeable. If the nationalized industries are to be made answerable to the public at all, there is no reason why they should be run by Parliament. They can far better be made the responsibility of those who are more or less directly concerned with them.

Liberal democracy requires then the arrangement of pressures on the individual so that the general good is not swamped nor the temptation to demand—and be given—too much by a supposedly beneficent Government elevated into an overriding rule of politics. It requires that democracy should not only express the will of the people but should see that that will is formed by a people

LIBERAL POLITICAL REFORMS

well trained in judgment, well-informed and furnished with a background of reason. Liberal democracy requires institutions which are grounded on a view of the functions of government. It rejects the equation of democracy with one man one vote regardless of the issues at stake or the circumstances under which the vote is to be cast. It believes that debate in a Parliament or discussion round a table require stage management. Above all, it regards the sphere of joint democratic action to be limited. The ultimate test of democracy is as much what we can achieve individually as what we can achieve together.

4

Liberal Economics

I have so far dealt with the political side of Liberalism. This is the side which has been neglected by Liberal politicians. Nor have British Socialists lately paid enough attention to political reform. Political progress altogether has got too little attention in the democratic countries. Yet very few of our economic decisions are, or can be, taken purely for economic reasons. From the politician's point of view the change from the art of political economy to the science of abstract economics has been a pity. Liberals look on economics as a means for increasing human happiness and creating prosperity in a humane world. They regard economics as a science, a servant, to be used in the running of a Liberal Society.

What sort of an economy then should a Liberal Society have? Political Liberalism has long been bound up with capitalist free enterprise, the competitive economy moved by profit. But we should not assume that the Liberal values can only exist under this system. Nor should we assume that the system is simple and obvious. Liberals seem to me to have spent too little time lately in examining the justification of the political economy associated with their name. They have pointed out the dangers of Socialism. But Socialism in its Western forms is not the only alternative. Indeed, in their present forms, Socialism and Free Enterprise share a great many of the same motives.

The British Labour Party have no objection to high rewards and big differentials. Their leaders draw substantial salaries and some are willing to serve private enterprise for large fees. These same leaders take all sorts of rewards in kind offered in political and commercial life. I am making no complaint about this: I am merely stating what is the custom, and a perfectly intelligible custom at that. Competition is not confined to business. There **is**

LIBERAL ECONOMICS

competition of one sort or another in most activities of life and it is not eliminated by Socialism. There is competition in politics: democratic Socialists approve of this. Even in business Socialists seem now resigned to the competitive capitalist system. The latest proposals of the British Labour Party are that the State should buy itself into business. In fact Socialists go further than acquiescence in the system. They want to run it.

Today, therefore, a Liberal must submit his beliefs in the private enterprise system to a more radical criticism than is now provided by the British Labour Party.

The Free Market permits a man to extort from his fellows as much as he can get, for his skill or his good fortune. It smiles on acquisition. Its working means that from time to time men are thrown out of employment or go bankrupt. In theory, at least, it keeps everyone in a state of struggle and uncertainty. Struggle and uncertainty cause unhappiness. As I believe that happiness is something which politicians should promote, I am not prepared to take the system on trust as obviously excellent. As to its moral basis, I hesitate to pose as a theologian, but I doubt if the free enterprise system in all its forms is always entirely in accord with Christian teaching.

Have we not reached a stage in world history when co-operation is more useful than competition? Instead of running our economy on self-interest, enlightened or otherwise, should we not try unselfishness? Could such an outlook be encouraged without concentrating power in the hands of the State or reducing the position of the individual? I am sympathetic to this point of view. The proposition must be examined.

Most men and women in different ways are anxious to be successful. They are anxious to succeed relatively as well as absolutely. But it is a valid criticism of modern life that it is aimlessly competitive. You only have to travel in the tube at rush hour to be caught up in the general struggle to get ahead whether there are trains to catch or not. It is extraordinary how much men and women will do for a little extra prestige, wealth or power. No doubt a great many people remain in comparatively poorly-paid jobs because they like the work. But it is also true that many leave their jobs or their country even in search of higher pay.

LIBERAL ECONOMICS

Where competitiveness has a worthy aim it is good. The man who wants to earn more to develop his own interests or those of his family is justified in trying to push ahead. The chief fault of the competitive system, as I see it now, is not that it leads to jungle warfare in which the weakest go to the wall, but that it is becoming an end in itself. So long as more goods are produced politicians are satisfied, regardless of whether they are the right goods or whether they greatly increase our happiness or well-being. We often hear phrases like 'Production has gone up 3 per cent'. Such phrases seem to me meaningless. I cannot understand how you measure the production of one pre-war Rolls Royce against the production of 500 post-war nylon shirts. But even if this is possible, why is extra production good? It is good to raise the material standard of living, certainly. But if in the process a growing number of people are relegated to work on endless belts—if in the process we create the same old slums or the new colonies of advertisement-ridden suburbanites, spending a tenth of their working lives rattling along in jam-packed trains, is it worth it?

The main defect of our way of life is nothing to do with the economic system at all. It lies in our failure to define the ends which the system is supposed to serve. As I have said, any system of political economy is a means to an end. The end is not merely the maximum production. When an American friend of mine told me in a discussion on nuclear weapons and possible disarmament that I must remember that to his fellow-countrymen it was unthinkable that you should use a technically inferior weapon if you knew of a superior one, it seemed to me that he was describing the climate of opinion in a lunatic asylum. Reduced to absurdity the argument leads to the conclusion that however disastrous to you or anyone else you must use nuclear bombs even in the smallest war. The same lunacy seems to have gripped the aviation industry. For reasons of national prestige every air-line is compelled to buy newer and faster aircraft year after year. This can only be done by drawing on a subsidy from the tax-payers and raising prices so that only expense-account travellers can fly. At the same time we also subsidize the makers of aircraft and engines and the airports. The upshot of all this is that the general taxpayer subscribes to the travelling comfort of the richest in the community

LIBERAL ECONOMICS

(while many country bus services serving the poor are in grave difficulties). Further, air travel gets more expensive and less convenient. Even on a slow transatlantic flight you are apt now to be thrown into a sleeping New York at about 5 a.m.

But these lunacies are not the fault of the free enterprise system. On the contrary, it is only by deliberately excluding independent operators from the air that we are enabled to keep up the present farce in international aviation.

There is, too, a perpetual clamour for more investment. It takes many forms. If an industry is doing badly, the popular but peculiar answer is to invest more in it. We are also urged to invest in the 'underdeveloped' countries while stepping up still further our investment in the developed countries. National greatness has become measured by statistics of investment. Further, there is a strong emotional bias in favour of big schemes and in favour of industrial investment. The erection of vast dams, factories or oil refineries is called imaginative. The investment in smaller aids to human happiness, in beautiful objects on a human scale or in simple and cheap creative work, all of which needs far more imagination, arouses little enthusiasm.

As an American economist[1] has said, 'The basic goals of economic policy should be part and parcel of the civilization of a Society. Ours are not. Our basic goals are the same as the goals of the Russians,' and Professor Stigler goes on to say that, 'Our very concept of the humane society is one in which individual man is permitted and incited to make the most of himself. The self-reliant, responsible, creative citizen . . . is the very foundation of democracy.' Right, but then what must we do?

The purpose of a high investment Liberal economy should be twofold: to expand the area of individual choice and to enable us to undertake more in the field of communal enterprise. This last is a vague term which I must try to explain further. It has always been recognized by Liberals that there are some chores which need to be done communally—refuse collection is an example often given. The field has recently been greatly extended. I do not think it has always expanded in the right direction but there is un-

---

[1] George O. Stigler, 'The Goals of Economic Policy'. Henry Simmons lectures at the University of Chicago.

doubtedly a field for public management as well as private enterprise: public management not always being exercised by the Government. Indeed, I think we have often got the alternative wrong: it is not primarily between Free Enterprise and Nationalization: it is between the tasks done for profit and those which are not.

Liberal values could be cultivated and Liberal choices made in a society of Gandhi-like self-abnegation. Though Liberals are not afraid of human choice, they equally recognize that choices are not made in a vacuum: that the free enterprise system does not of itself or of necessity improve the circumstances in which choices are made and that Liberalism is not committed to its wholesale operation. The Western world is, however, committed to a high material standard of well-being. This standard cannot be maintained with political freedom except by the use of the free market system. But the Free Countries must turn their attention to educating their children in putting their good fortune to good use, and there are some lessons which Liberal politicians can learn and apply from a more unselfish system.

It must deal kindly, for instance, with those who want to contract out. I deal with the Social Services separately. They have become a way of helping those whom the system has treated harshly. But there are always some people, people of value to a civilization, who want to keep clear of the System, who don't want to compete at all. And by compete I reiterate that I do not mean only compete for money. There have always been in civilized societies places like the church, the universities, and some branches of medicine, open to non-competitive people. We are in danger, I think, of carrying the race for promotion, business, wealth, into unsuitable fields. I rather shrink from the phrase 'Opportunity State'. Opportunity by all means; but the Opportunity State is coming near to being a contradiction. The State's function should not be to provide a means for the getting of material gains, it should not be a vehicle for that sort of ambition: for when the State competes it becomes entangled with every sort of ambition for prestige, it becomes tyrannical, and into the bargain it is usually inefficient. There is a better role for the State in protecting public bodies which can offer different satisfactions. Personally, I would hope

LIBERAL ECONOMICS

that a Liberal Government would go far in fostering certain non-profit-making activities. I deal with its role towards the arts later. I argue that a Liberal cannot be satisfied with the present position over town and country planning—and that this is a matter where free commercial enterprise is unsatisfactory. I am not averse to the help now given to churches, especially for the upkeep of their buildings. There are certain crafts which even if their results are not economically important ought to be State-aided. If politicians are going to drive on to higher production and take credit when new records are reached, they should also set a value on more leisure. Liberals should devise schemes whereby business men could change places with Civil Servants. A Liberal Government would be serving the true ends of a humane society by encouraging 'sabbatical years' in the Civil Service, trade and industry. Certainly roads, housing, education, should all be maintained at a high standard even if this interferes with Free Enterprise.

Over most of the field of production, distribution and exchange, however, Liberals believe in the Free Enterprise system. Their reasons are, broadly speaking, threefold. It is in itself a freedom. Of course, it is true that you are only free to go into the Ritz if you can pay. Of course, many people have dull and comparatively ill-paid jobs which they could only leave with difficulty. But freedom to spend your money, change your job, save and set up your own business is nevertheless a freedom of importance. Secondly, the system can be more efficient than any other. Thirdly, it can be made more acceptable to more people than any other system.

The ends which the free system serves have been distorted by politicians using the machinery of government. The same holds true of many cases where it is said that the system rewards the wrong people. The take-over bidder is in the market because the restriction of dividends has depressed the true value of a company's shares. The capital gainer of all types is in the business because high taxation makes a capital gain so much more worth while. Restriction and rationing breed a black market. So the story goes on. If the State makes too much use of the system to serve its own ends of national prestige or 'social' justice it promotes those very vices which Socialists most decry.

The Free Enterprise system is fundamentally unsuitable for a people which wants its Government to make up its mind for it. It is suitable for a people who are prepared to decide themselves what they want and what is right or wrong. Liberals believe that these choices can be and should be left to individuals. They are not therefore put off by the more superficial criticisms of Free Enterprise. Nor are they frightened because some people will make what most people will think are the wrong choices.

The system is naturally associated with private property. I do not myself hold the view maintained by most Liberals that private property is inextricably bound up with Liberalism. I grant it in the extreme cases of tooth-brushes and false teeth, which really cannot be held communally: but I can imagine Liberalism flourishing in a Kibbutz—in fact, the State of Israel has a great deal to teach Liberals. But if we want high industrial production and an expanding society we must either accept a very great concentration of power in the hands of our rulers or find new ways of spreading property. And I believe that the concentration of economic power in the hands of the State is a threat to Freedom and Liberalism. Property is a bulwark against tyranny if spread sufficiently widely. The practical choice before us is to spread it or see our freedoms curtailed.

For Liberals, therefore, the Free Enterprise system associated with private property seems the best instrument to hand and they intend to make it work in a liberal fashion. But we emphasize that Free Enterprise is a system, an artificial system, which can be changed. It is not, as it is often represented, a state of anarchy. Liberals would reform the present system.

The free competitive system is an artificial way of carrying on economic life. It depends, for instance, a great deal on the concept of limited liability. Without this we should be forced either to give some other shelter to enterprise, probably by getting the State to undertake most of it, or protect it by charters, or we should have to forego the rate of advance we have enjoyed over the last hundred years. But the law relating to limited liability is not the only benefit conferred on private industry by the community, there are such statutes as the copyright and patent laws and, above all, the fundamental structure by which contracts can be enforced and law and order preserved.

LIBERAL ECONOMICS

Limited liability and the Company Acts are of the greatest importance to Free Enterprise and must be kept up to date.

Apart from spreading ownership Liberals would change the Company laws to give workers status.

The conceptions of trusteeship such as exist in the Lewis Partnership and the Carl Zeiss Foundation are in line with the development of the Pluralist Society and Liberal free enterprise so long as they are associated with competition between Trusts, so that the consumer interest is still ultimately protected. The appropriate body should not only have its powers to check monopolies strengthened, but might also be given authority to stop certain practices which lead ultimately to monopolies, such as the deliberate war on a competitor by selling under cost price in his market alone and some of the practices associated with take-over bids. Further, the types of State enterprise which would be most beneficial are never suggested by Socialists. There may be a case on empirical grounds for the State taking over, or starting anew, a factory, shop or service where private enterprise is suspected of failing to give satisfactory service to either consumers or producers. There have been from time to time suspicions that private cartels have delayed the exploitation of new inventions which would harm their business. This is a field where State intervention might serve consumers well. It might have been much more effective if the State had nationalized only a few coal mines or a single coalfield.

Within industry one of the most difficult questions is the reward to be offered to workers. Co-ownership is, Liberals believe, the key to this. But co-ownership, though it will make it easier to determine wages, will not of itself solve the difficulties in wage negotiation. As Lady Wootton and others have pointed out, there is no rhyme or reason in the present methods of settling wages. This might not matter to politicians had there not grown up since the war a widespread belief that wages were one of the main villains of the 'cost-push' inflation from which we are said to have suffered. The Unions have been blamed for the loss of value in our money. No doubt there has from time to time been a cost-push inflation—though 'cost-push' must quickly pass to 'demand-pull' in the absence of new factors. But 'cost-push' or 'demand-pull' can only succeed when the amount of money or credit necessary to sustain the inflation is made available by the Government. Most of the time since the war it has been the bidding of employers for labour and their ability to pass on extra wage-costs which has been the mainspring of the inflation.

Liberals dissent from the jargon about wages which has been bandied about in the post-war world. The business of Unions is to drive a good wage bargain for their members. If the members of these Unions felt themselves to be partners in industry this would be in itself a powerful check on the bargains being too hard. For a powerful union to make such exorbitant demands that employers were driven into bankruptcy would be suicidal from its point of view if the workers were themselves largely interested as owners. But to ask Unions to moderate their demands below what employers are willing to concede because there is inflation is to cure the symptom and not the cause. It is for the Government in the first place to curb currency, credit and its own spending. The Unions may have been guilty of restricting output but I doubt if they can be blamed for causing inflation, except at one remove back. Inflation is a political problem. Its cure is not so very difficult but it is often politically unpopular. Like smoking or drink, habits in which everyone recognizes the evil of even mild excess, except in their own cases, so everyone is keen to denounce higher prices except for their own products or labour. In this sense the Unions may contribute to inflation: as a well-organized producer interest they have made their weight felt against deflationary policies.

But though I am not much impressed by the argument that the Government must concern itself over wages because wage-rates can cause inflation, I think the reward to workers in industry is certainly of political importance. It is of importance that there should be a fair reward. There are still pockets of employment where exploitation is not impossible. The case for legal minimum wage-rates is much weaker than it was, but it still exists, even if only as a long-stop, a guarantee that exploitation will not be permitted, something like the reserve at the Bank of England which gives a feeling of security though it is never used. Apart from the minimum rates, however, the determination of actual earnings is

LIBERAL ECONOMICS

a matter of great significance in political economy. It can give rise to friction, it can be very disruptive of a society if it is felt to be unfair. The sharing of profits in a Liberal Society will remove one source of irritation. Though profits are not ordinarily a cost like wages, high profits going into a few hands are a source of irritation when wages cannot be raised generally and high profits, after all, are often an indication that price reductions which could be made are not being made.

A Liberal policy about wages stems from its belief that the Free Enterprise system properly used is a willing horse in harness and not a mad bull. Industries should be left to settle their wages in the market. The employer offers what the job is worth to him and the employee sticks him up as high as he can. If the profitability of the job increases the employee should ask for higher wages. Liberals see no reason why this should not be normal practice. But having established the normal practice they admit that it needs certain qualifications.

In line with the Liberal view that the area of Government action should be reduced, but within that area government should be more effective, we believe that the Government has a responsibility for the regulation of credit from which it must not abdicate. We believe that it is primarily through proper financial management that inflation or deflation are to be kept within bounds. We do not, therefore, favour the idea that the T.U.C. or the Employers' Federations should be consulted from time to time, the blanket policy for wages agreed upon, and Unions and Employers then told that they can haggle over the 3 per cent or 5 per cent or whatever it is decided the economy can stand. We are not converted to the Swedish method of wage negotiation. But equally Liberals who all want a Pluralist State believe that consultations with Union leaders from time to time about the general state of the economy may be valuable, chiefly as a means of making the facts of the situation generally known and encouraging the Unions to press for industrial efficiency so that earnings can be raised. The consequences of various choices about wages must be widely understood and if the Unions, or for that matter the Cohen Council, can get these over to the public, so much the better.

A wages-market operated by supply and demand and related

to productivity, which should be the normal situation according to Liberal thinking, must be subject to two formidable limitations. Productivity may be meaningful in some jobs: it is impossible to assess it in others. Nevertheless, some gauge of productivity lies in the profitability of the industry as a whole. Workers in jobs where the direct contribution cannot be assessed should share in proportion to the increased profits made by their industry where their jobs continue to be essential to it. In jobs such as the Civil Service where productivity and profitability have little or no relevance, even taking the service as a whole, wage increases should be negotiated with an eye on any overall increase in productivity. But as this type of job usually has many advantages, such as prestige, premium rights, security, it should not expect to gain so directly from increased productivity as do industrial jobs. Indeed, while over the whole range of employment wages should rise with productivity, the amount they rise must be related to the industry or the firm or indeed the job, the more productive paying more. There will also be in the longer term the factor of the supply of various types of labour.

The picture which Liberals envisage of more piecemeal negotiations over and above a basic minimum may not be so very different from what happens now. But there will be more onus put on management to manage. Further, if the national minimum alone is agreed at the summit, a great deal of friction and posturing for prestige reasons will be eliminated from wage negotiation. Such a pattern of negotiation, broken down to discussion of the worth of various jobs and the profitability of various industries, will need a greater dispersal of knowledge of the economic situation and a greater willingness to take decisions at a lower level and accept more differentiation—all of which Liberals should welcome.

These objectives to be aimed at in assessing the reward to labour should be considered with some warnings against misunderstanding. Because it is agreed that extra profitability should lead to extra wages this does not mean that I assume it will always be due to extra effort on the part of the workers. It may be. If so, it will probably be in the type of industry in which piecework (which despite current prejudice against it has much to be said in its favour) is appropriate. Where this is in force extra reward to those making the chief contribution may be automatic. On the other hand, the extra profitability may come from some new invention or the opening up of a new market to which most of the workers have contributed nothing. If they are substantial owners of the business they will reap their rewards through extra profits. But if their share in the ownership is not great they should still see some addition to their wages as people who are in a limited sense partners in the business. But the increase may well be small. Nor must it be thought that Liberals assume that wages will necessarily fall as unemployment increases or profitability decreases. In the period after 1918, when unemployment was high and many industries were on short time, wages not only showed their classical resistance to reduction but actually increased by as much as 4 per cent in some years. This was at a time when wages had lagged behind. Now the wage earner is much more fairly rewarded. However, while wages can hardly expect to continue to rise against decreasing productivity, it is desirable that, as they will not rise commensurately with increases they should resist commensurate decreases in more difficult times.

The difficulty of fixing wages in nationalized industries is one sign of the unsuitability of nationalization as a method of running industry. So long as the State stands behind an industry, the criterion of profitability disappears. It may be written into a nationalization act that the industry is to make ends meet taking one year with another. But it is a Canute-like edict because there is no sanction behind it. The Government eventually pays. As most of the nationalized industries are monopolies there are few comparisons and little commercial competitive pressure towards efficiency. So long as they remain with us in their present state, wages in the majority of nationalized industries must either conform to those paid for similar jobs elsewhere or, where there is no equivalent, be fixed on Civil Service lines. But Liberals intend to alter radically the management of these industries.

The rewards to management have received less attention than either wages or profits. As I have said, provided that they are widely shared there is no strong objection to a large distribution of profits in an industry where nevertheless wage-rates rise much less quickly. But rewards to management are a cost of exactly the same type as wages; if the workers are to accept negotiation of wage rates on a narrower front, and if they are to be given a more responsible status in industry, salaries, expenses and other emoluments should be openly stated, and should also be subject to negotiation in the light of productivity. It is difficult to assess how much the rewards of all types, including expenses, which are now paid to management, have risen, but a study of directors' fees in company reports shows that they have risen very steeply, probably more steeply than wage rates. Private ownership widely spread, high wages and a well-organized market, these are fundamentals of a Liberal free economy.

But a free market in anything saleable is not so easily achieved as is often assumed. Markets have to be organized as they are in the city. Here, indeed, is a field in which the Government may intervene to improve the Free Enterprise system. It would often be much better employed creating markets than destroying them. There is always a tendency towards monopoly. This is partly due to recent State action; the tendency for Government help and contracts to go to big firms, for instance, and heavy taxation which has often favoured the big established company. But there are other causes of this tendency. There is the efficiency in some circumstances of large-scale operation: there is the need for research beyond the scope of small companies: and there is the quest for security. A situation of continual and rabid competition, though it may suit the consumer and the successful producer, creates uncertainty and unhappiness which a Liberal must set against any redeeming benefits. Liberals should not be dogmatic about the size of industrial units. What they should guard against is the growth of the type of monopoly which restricts entry into an industry or trade or the development of substitutes. They should remind the public that free imports are one of the best antidotes to exploitation by monopolists and they should seek to improve and change the powers of the Monopolies Commission and the Restrictive Practices Court. The purpose of restraining monopolies and restrictive practices should be twofold, to protect the buyer and to ensure that our resources can be redeveloped continually to their best advantage. Resale price maintenance is not compatible with the former, nor restrictive industrial or trade practices with the latter, but an attempt to break up some of the big amalgamations such as I.C.I. might lead rather to waste than full exploitation of our resources.

To keep our resources fully employed on the tasks which are either the chosen goals of individuals or are the accepted aims of the community must, as I have said, be a major goal of economic policy. This may from time to time mean the direction of some of our effort to an end chosen by the community. Dynamic societies are usually working to some such end, usually, unfortunately, nationalistic. Britain, at present, could well devote a proportion of her resources to helping less prosperous areas, or to such communal undertakings as roads, slum clearance or the building of universities if her output was higher. A combination of private and public aims is what Liberals should develop.

This brings us to the overall financial policies which Liberals should pursue. In particular, we must consider whether we should aim at rising, stable or falling prices and whether we should accept a greater degree of unemployment for the sake of lower prices.

It is fashionable to speak as though inflation and rising prices are necessary to keep the economy buoyant. It is significant that this is a line of thought frequently pursued by Socialists. It is not historically true. Productivity rose at times between the wars in spite of stable or falling prices and a generally deflationary situation. If it were true that stable prices were incompatible with full employment and rising productivity this would be a most potent argument for re-examining the whole of our financial structure. We have not, in fact, examined it closely enough—and it does need considerable change—but it is the present type of State Socialism which has made inflation appear necessary and even desirable. That 33 per cent of the national income is taken and spent by public bodies is in itself inflationary, as has been shown by Mr. Colin Clark. The diversion of large blocks of wealth to investment in the nationalized industries is inflationary. By definition this investment is relatively uneconomic. The argument for it is not that it will be profitable but that it is needed for some other reason. It will nearly always fail to add as much to our material wealth as it would if invested in some more profitable way. But it can be made to appear desirable so long as inflation continues. Inflation is both encouraged by Socialism and is a cover behind which the true economic failure of many Socialist schemes can be camouflaged. It is made worse by Socialism associated with democracy because interest groups continually press the Government down the path of further inflation and a Socialist Government is always keen to take on more, to spend more and to expand its activities.

I suspect that lower taxation and the alteration of the present form of nationalization, as well as a new political outlook on the function of government, are needed if stable money is not to be associated with slack resources. One of the most distressing features of our post-war economy is the penalty on earnings. Win a football pool, carry out a deal on the Stock Exchange, inherit some money and you will gain far more than can a professor, a doctor, manager or a technician from his salary. We must make it more worth while to earn an income. That means reducing taxation. By both rewarding more generously the man who is living by his productive skill and allowing him to spend or save more of his earnings himself you will make the economic system more stable and less subject to gusts of optimism or pessimism raised by politicians.

Liberals do not therefore concede that inflation is essential to full employment. Nor do they consider inflation desirable. There is a school of thought which holds that as inflation merely transfers wealth from some pockets to others it is not as objectionable as is sometimes thought. But inflation makes the balance of payments problem unmanageable. It is incompatible with sterling as an international currency. Worst of all, if anything is anti-social, it is. It is notorious that lenders, pensioners, salary earners, those in weakly organized Unions suffer from inflation. Small firms find it difficult to finance replacements of capital. Speculators and large firms gain. Without claiming that the group who suffer from inflation are morally superior, they are clearly a group who should not be discriminated against. But that is what inflation does. To try to counteract it by higher pensions or subsidies is never effective, nor is it desirable, for as a general rule it is much more desirable that people should not be subsidized than that they should.

I see no reason why in a Liberal economy without inflation unemployment should rise above the Beveridge level. We are obsessed with the pre-war type of unemployment. This is a legacy from the 30's and a sign of the conservatism which is deeply rooted in our politics and to which the Welfare outlook contributes. We should be concerned with the full use of our resources. We should welcome change if it means more leisure for men and fuller employment of machines. But in industries such as cotton we do not employ machinery to the full.

General slackness in our resources can now be taken up on Keynesian lines if we have international co-operation and a supranational foundation to our economic and credit systems. Unemployment in particular industries or districts should be met by specialized Government measures. The direction of industry towards certain areas is justified as I shall argue later, not only on economic grounds but as a measure of Town and Country Planning. No doubt, too, the Government should try to regulate its public works programmes so that they can be accelerated in slack times. I do not, however, feel that this can have much more than marginal effect. The amount of labour required on, say, road building, has diminished. Even when the schemes might give significant employment, most public works cannot be put forward or back to suit overall economic conditions.

But that the Government should get away from annual budgeting and annual Treasury control so that in the planning of credit and financial policy it is not tied to the calendar seems very desirable. 'Autumn Budgets' are becoming a familiar idea. But they are associated with crisis. It may be convenient to keep the traditional budget as an annual review of how things are going and an annual examination of our taxes. But for many planning purposes the Treasury should tear itself away from the April to April mentality.

We should make use of flexible cash ratios and for some purposes of Treasury Deposit Receipts. There seems little to be said for the system by which Government borrowing through Treasury Bills automatically inflates the credit base. The Government must finance more of its needs by funded loans: but apart from this the credit position should be regulated by considerations other than the Government's needs.

We cannot treat Britain's economy in isolation. The most

serious aspects of our present economic problems are international. Just as the cure of unemployment in pre-war days would have needed the abandonment of vested ideas as well as vested interests, so it is today. There is a lot of Schachtism about. There is a lot of economic nationalism. A feature of current economic national selfishness is that it often relies on other countries, particularly America, to practise national unselfishness. A great deal does depend upon the Americans, but if we are to expect them to show international virtue we must not hug our own national vices. Many countries have suffered from a more or less chronic shortage of dollars since the war: Britain has had recurrent troubles over her balance of payments and the exchanges of the Free World have often been in crisis. The serious national aspect of this is that it greatly limits our expansion and indeed our room for political as well as economic manœuvre, while it might prevent the Keynesian remedy for a slump because our reserves might prove inadequate to meet any lowering of interest rates.

A common reaction to this is to run for cover behind controls of one sort or another such as control over the movement of capital or controls on certain types of import. There are many who argue that it is the preponderance of the dollar coupled with the comparative unimportance to Americans of their foreign trade, which, so the argument runs, puts the non-dollar area at a permanent disadvantage. The obvious remedy appears to these people to be careful discrimination in what we buy for dollars and the development of other sources of supply. Liberals cannot ignore the difficulties, psychological and economic, of the post-war world, but they cannot accept this solution. It involves the kind of restriction on choice to which a Liberal must be opposed except as a temporary crisis move. It puts a brake on the proper deployment of resources. It will also play into the hands of the Communist bloc. For it must be a disruptive move. If every country attempts to safeguard its reserves and improve its balance of payments by restrictions it must lead to the general weakening of the Western world. The Communist world with its centralized direction may soon be in a position to carry on economic warfare of a very effective kind if the Western economies are not strengthened and united. Liberals must reject unity based on dirigism of the Com-

LIBERAL ECONOMICS

munist kind: but national dirigism based on protection would not even have the advantage of international dirigism which at present is in its full form impossible within the West.

From every point of view, therefore, it is essential that unity should come to the Free World—and it must come through freedom and expansion. For this, two things are necessary. Individual nations must curb inflation. In Britain less government spending—co-ownership, a sensible wages policy, monetary discipline and improved monetary techniques all have their part to play. At the same time the reserves of the world must be increased. World reserves of gold have not nearly kept up with the increase in international trade. We should be in an even more serious situation than we are had not foreign balances made up some of the difference. But even so the reserves are inadequate. Furthermore, the reliance on foreign balances is precarious. Unless reserves can be increased, world trade must be hampered and in particular that very large segment of trade which is transacted through sterling. The difficulty is to increase the backing of such currencies as sterling so that full convertibility can be reached soon without starting again on inflation. Dollar backing for sterling would involve a very large figure in dollars. To increase the price of gold, coupled with increased quotas to the I.M.F. seems the simplest remedy: unfortunately, it does not appeal to the U.S.A. This is not the place to state dogmatically which remedy must be applied. The need, however, is to back British expertise in international finance and the markets which exist in London with adequate reserves which can only at present come from America. The Sterling Area is often attacked, usually because it is said to cost us more than it is worth. The main count against it is surely that if it worked ideally it would not be necessary. In one form or another, however, it goes back for some time and its sufficient justification is that it is a very useful piece of mechanism in the present set-up of international finance. The benefits which it confers on Britain and many other countries could not at present be got in any other way. But if British world banking is to continue it is perfectly true to say that it must have support, particularly from America. It is very necessary, too, that America should liberalize her own policies and reduce her own tariffs and restrictions.

LIBERAL ECONOMICS

Britain must continually press for the widest area of freedom and economic unity. The failure of British Governments since the war to take the lead in time when there were possibilities of uniting Europe has been consistently criticized by Liberals.

To sketch again the general background of Liberal economic thinking: We regard economics as a means only. We concede that there might be an economic system totally different from that of Free Enterprise which could support a Liberal society. But at present such a system could probably only do so if we accepted a lower general standard of life. The Liberal instrument, therefore, is the Free Enterprise economy coupled to private property. But this system must have worthy aims and it must be realized that it is a system subject to control and alteration. Its aim should be to increase the prosperity and freedom of choice of individuals, leaving a margin over for the support of communal activities. These activities need not be profit making. Indeed, the failure of nationalization has been partly the failure to realize that if something is best done or run by the State it cannot be run on free enterprise, free market conditions. The attempts to make nationalized industries run as private industries when they are in a different position over the raising of their capital, the fixing of their wage rates and the pricing of their products merely leads to muddle. I attach great importance to lower taxation, to the wider distribution of property, the methods of fixing wages and salaries and the financial control not only of our own economy but of the free world. They must set aims for the economic system and want to reform its methods. In the long run aspects of the economy will fall into place as Liberal political and economic reform proceeds. But, in the meantime, there are certain aspects of our present economy on which the Liberal view needs to be restated.

The first of these is Free Trade. As far as trade is concerned, Liberals want to see it flow with as little impediment as possible. I would not, however, completely rule out any tariff or quota at any time. If you concede the need for planning at all, these measures may in some circumstances be a legitimate planning instrument. The norm is free trade. Deviations may be justified but the onus of proof lies very heavily on those who want to deviate from the norm.

LIBERAL ECONOMICS

There seems to me one clear case in which the Nation State is justified of its own volition in imposing import restrictions. That is when another State virtually declares war on it by deliberately selling goods in its markets far below the economically justifiable price, not for trading reasons, but because it wants to smash the whole economy or certain industries of the recipient country. If Russia turned the whole resources of her all-powerful State onto disrupting the markets of the free world Liberals must treat this as an attack on freedom, if necessary by resorting to the type of economic 'siege' measures which are otherwise only acceptable during actual war: though the methods of defence might certainly be more subtle than the imposition of an embargo or a very high tariff. The Free World should gratefully receive all, or at least a high proportion, of the cheap imports offered and devise means of sending them to the poorer people and countries. But individual Governments would be entitled to take steps to prevent the economies of their countries being disrupted.

When it comes to dumping, short of economic war, Liberals would like to see this tackled internationally. Governments which subsidize exports which are then sold below cost price are committing an international crime unless there is some special reason for their action. Both subsidization of exports and restriction of imports may cause hardship and in most cases do more harm than good. Here is a field in which states should yield some sovereignty to supranational bodies which have some power to regulate imports and exports in cases of dire need. Already G.A.T.T. goes a little way in this direction. But it is not a supranational or executive arrangement, and it assumes, not free trade, but a degree of protection. The British Government may have a case for regulating some types of textile imports; if so, this case should be made to an international body. Agricultural and horticultural products are peculiarly difficult to market: the sudden appearance of Argentine beef in great quantities can cause great distress to British farmers: an attempt to help, say, New Zealand may cause distress in Denmark or Holland. Here is another field in which some international regulation is needed. Until a responsible body is in being which can do this sort of work, Government subsidized dumping, even if it is done not for reasons amounting to economic war but in an attempt to get over a temporary difficulty, is something against which other Governments may have to take action. But Britain should use all her influence to get proper supranational control of trade when the ideal of free trade must be departed from.

The argument for free trade depends upon all the rules of the game being observed. For instance, the division of labour must be accepted so that goods are produced where they can be made most cheaply. But country after country today is building up manufacturing industry without much regard to its economic justification. If a country sinks £700 million in creating a steel industry and trains numbers of workers for that industry it will not be keen to see it put out of business by cheaper steel imports from countries more favourably placed. Nor is the world's financial system favourable to Free Trade. The simple theory was that if a country imported more than it paid for by exports, it lost gold, its prices fell, it could not afford more imports while other countries acquired gold and became more profitable markets. But America does not play the game according to pure theory and its predominance in the economy of the Free World raises very real problems. There is, too, the emergence of the Sterling Area and of currencies managed for political reasons.

If the solution to World Trade difficulties is a supranational body, individual countries must also give up their absolute right to manipulate their economies and finances regardless of the effect on their neighbours. But this, too, is an ideal. In the meantime Liberals should press for the development within the Commonwealth, N.A.T.O., and Europe, of general agreements relating to the control of international finance and the development of industry. The countries which supply capital for development in underdeveloped countries could well insist on the creation of an authority to check the creation of industries which can only work behind high tariff barriers and with subsidies for all time. It is important that any such check should not be absolute, for there may be areas in which uneconomic industries will nevertheless be essential, or there may be types of industry which it is necessary to keep in certain countries even with Government aid. But these should be the exceptions. At present the deliberate encouragement of uneconomic industries is almost the rule, even in countries too

LIBERAL ECONOMICS

small to be able in any circumstances to support the full gamut of industrial life. More important still, any such check should not be imposed by the lending countries as a condition of their loans. It should be decided upon by a supranational body or, until such a body can be created and made acceptable, existing institutions such as the World Bank—through which Liberals would like to see more loans for development channelled—should give advice and when necessary exert pressure in the cause of international planning. The reverse of this process should be the encouragement of joint efforts by the more powerful industrial countries to help the weaker, such as the setting up of a consortium for a particular task as has lately become common.

Another corner of the economic scene which needs the attention of Liberals is the national boards which do not run industries but exercise powerful control over them.

In the last twenty-five years Governments, both Tory and Labour, have contracted the habit of setting up advisory bodies, or regulatory bodies, for industries which are in difficulty. There are the various marketing boards, there are the White Fish Authority and the Herring Industry Board, there is the Forestry Commission. In addition, there are bodies like the Local Health Executives which are loosely allied to the local authorities. These bodies, from a Liberal point of view, suffer from several faults. They are often irresponsible and frequently wield monopolistic and absolute powers. The sanction of complete boycott which may be effective even against a private monopoly is of no effect against them because it makes little or no difference to their members, who often regard themselves as Civil Servants and whose salaries depend not on the economic success of their undertaking but on the Government's policy. To be effective they should be development bodies with a commercial outlook: but they behave like Government departments and are filled with bureaucrats actual or in spirit. They are not in practice responsible to Parliament, and Ministers themselves sometimes propound the extraordinary doctrine that they cannot control them, so grievances must be taken up direct or through the ineffective consumer councils who work hand in glove with the boards, often sharing some of the same members.

LIBERAL ECONOMICS

Many of the boards should be co-operatives run by the people, consumers or producers who stand to gain or lose by their activities, and working in conjunction with national or regional governments. For instance, there should be a Highland Development Authority composed of representatives from farmers, fishermen, weavers, etc., co-operatives in the North, together with representatives from a Scottish Government. It should be thrusting and commercial in outlook and free of red tape. I have already mentioned that there are other fields, such as the Health Service, in which those interested, laymen as well as doctors, should have direct representation. By such methods you may get democracy of the kind which emerges when a few people who are interested in a particular object get round a table.

If once we have suitable democratic custom and institutions we can begin to deal with those subjects which lie outside the market.

As far as nationalized industries proper are concerned the few that might be kept nationalized at all should not be wholly nationalized. By that I mean not only that where possible individual units be left in private hands, but where one whole industry is nationalized alternative products at any rate should not be. Responsibility for many of them should be with those who work in them subject to overall control, but either units in the industry should compete or consumers should be associated with management.

We come next to Agriculture and Fisheries which, although they are not nationalized, are exempted to some extent from the full force of the market. One Liberal argument for giving them some support is that they contribute to a reasonable balance between urban and country populations. In fact, Liberals should have a 'countryside' rather than an 'agricultural' policy. The herding of more and more people into cities is wasteful. It is made possible by policies which encourage it. It can breed that proletarian outlook which is antipathetic to Liberalism. To make the best use of our resources for the building of a Liberal Society we need to have a healthy countryside embracing healthy towns and villages. The agriculture and fishing as well as some other small rural industries suffer from two peculiarities which cannot be removed entirely and which make the working of a free market difficult. There is the irregularity of supply. Beef, fish and vegetables are all apt to be 'dumped' without warning at a low price regardless of the market. There is also the great number of small producers. To forbid these producers to join together to enforce a selling policy, advantageous from their point of view, is as illiberal as to criticize workmen for joining unions to get a higher price for their labour. Much agricultural produce, as well as fish and vegetables, has to be disposed of by the producers or catchers within a very limited period.

These characteristics point to the need of co-operatives, not necessarily all-embracing, and certainly allowing some genuine competition between foods. The elasticity of demand for some products is small, but a badly run co-operative will drive consumers from its products if they have better value in other foods. I see no reason why regional bodies or national Governments should not be represented on these co-operatives. But one effect of trying to run agriculture too much from the central bureaucracy is that the co-operative movement in British farming has languished.

The Fatstock Marketing Corporation which was only set up when Government policy altered is a step in the right direction. It is to be hoped that it will be the forerunner of other attempts by the Farmers' Union to improve their industry. For the time being, at any rate, the uncertainties of agricultural production will entail some price support. Liberals should aim at stopping the dumping of surpluses, by international action, and at improving processing and storage of perishable commodities. A proper use of rent to ensure the best use of land and a supply of capital for farmers and fishermen are also needed. But this last should also be used to assist small-scale industry. Already in parts of England there is a tendency for some manufacturers (e.g. shoes and cotton) to contract out some processes to small units in rural areas. This is essential if the population in the Scottish highlands is to be held. It is quite illogical to subsidize and assist agriculture and fishing alone among rural industries. A sound agricultural policy must take into account the differing needs of different parts of the country. That is one reason why Scottish agriculture, for instance, should be largely within the province of a Scottish Government.

As for Land Tenure, owner occupation is the ideal. There is

nothing inherently uneconomic in small farms so long as the farmer is efficient and marketing and research are carried on by bigger co-operative units. But the perpetuation of very small crofts is a hopeless policy. Nowhere is the conservative tendency of the Welfare State more clearly seen than in agriculture. The Crofters Commission is trying to perpetuate a dying form of tenure instead of developing a new one. The freezing of tenancies has made it very difficult for energetic farm servants to acquire or enlarge holdings: while the difference between the price of land with or without occupation is a pointer to the amount by which agriculture is unnecessarily or wrongly subsidized. There are, too, parts of the country where large-scale schemes of development, e.g. land reclamation or drainage, are needed beyond the bounds of voluntary co-operation alone. Here is another field for public enterprise, through regional or national bodies.

These are specialized situations which Liberals face today. They may change or disappear as Liberal Society develops.

I have departed rather in these last pages from my general principle of not dealing with short-term or particular matters because Free Trade, the National Boards and industries, and Agriculture are so much in the forefront of many people's thinking today. I want, therefore, to finish this chapter by stressing that Liberals are more concerned about the general overall aims and methods of our economy than they are about particular problems which have largely been created by the departure of ourselves or the world from these general aims and methods with which I have dealt earlier.

While these general aims and methods will remain valid, Liberals must not allow themselves to be frozen in a set attitude towards the free enterprise system. Industry is changing all the time. If we are to take full advantage of the new sources of energy and the synthetic raw materials which may be offered by science, our industrial organization and financial techniques will need a continual flow of men like Robert Owen and Lord Keynes.

5

Co-ownership

The reasons which lead the Liberal Party to campaign for the spread of ownership are political or social as well as economic. We believe that the possession of some property is essential if a man is to enjoy full liberty. I do not want to misuse the word 'liberty'. I am not saying that everyone who does not have a house of his own or some savings is a slave. Nor am I equating the possession of property with freedom. But a certain amount of elbow room: a certain cushion against economic setbacks such as unemployment, are essential to full liberty and can only be possessed by those who are not totally dependent on the charity of the State or their weekly wage. The possession of some property widens a man's choice and gives him more scope to exercise his talents. Personal ownership is one badge of a citizen as against a proletarian. It is a shield against petty tyranny. Further, you cannot have a contented, industrious and stable society where property is largely concentrated in a few hands, or worse still under the control of the State. As has been said, 'In respect of property as opposed to income, Liberals are outright distributists.' In this we differ from the Tories who, whatever some of them may say from time to time, have in practice been content to see property concentrated in too few hands. It has been calculated that, whereas in 1911–13 the richest 5 per cent of the nation owned 87 per cent of all private capital, in 1947 the same percentage still owned about 72 per cent. If this underestimates the amount of 'consumer-durables' in the hands of the poor, television sets and so forth, it shows how great an inequality still persists. As for the Labour Party, though the methods by which they would do it are varied from time to time, they always want to suck private property into the maw of the State.

CO-OWNERSHIP

The maldistribution of property leads to friction between the 'haves' and 'have-nots'. Society is also pushed and pulled by the supposedly differing interests of organized capital and management on the one side, and of organized labour on the other. The relations of these two groups are too often described in terms of warfare. The stresses set up hinder economic advance, and the battle, when joined, is carried over into the political arena. As the Labour Party is largely financed by the Unions, it is not surprising that it often speaks as their political mouthpiece. The same is true, to a lesser extent perhaps, but nevertheless all too true, of the relationship of employers and the Conservative Party.

Yet this political-economic setup which is so damaging is neither inevitable nor does it, in fact, correspond to the realities of modern industry. No longer need we recognize the inevitability of classes distinguished by the kind of property they own, or the type of income which they draw. Certainly the simple idea of industry being a matter of workers on the one side and bosses on the other, has long since ceased to correspond to the facts of life. The owners, in the sense of the shareholders, have lost a great deal of their power. The managers have gained in importance. 'Workers' belong to all ranges of income. The Government has an important, sometimes a determining, role in industrial policy. But the old habits persist and we have not given nearly enough thought as to the way in which we are to run a modern industrial system.

The system ought to stress the community of interests among those taking part in production. It ought also to emphasize joint responsibility. The community of interests is plainly too often ignored, as anyone can see who reads the accounts of industrial disputes. The blurring of responsibility is not so obvious. But it is notorious that the shareholders, having put their money at risk in an enterprise, all too often take little further interest in its affairs unless they go wrong. Far be it from me to minimize the importance of risk-bearing, even in a limited liability company. Nor would I write-off the residual powers of equity holders. Though in the background most of the time, their shadow is always there ready to materialize into something more potent if management fails in its job. From time to time Shareholders' Committees are formed to assert the rights of the owners. Even

CO-OWNERSHIP

more potent is the vigilance of financial journalists whom it is profitable for newspapers to employ just because a body of investors exists. But in spite of this, the present Company Acts attribute to shareholders a role they do not fully discharge—and give them very considerable privileges on this false basis.

To Liberals, the splitting of ownership from management is not welcome. It is being accentuated through the repeal by many companies of the provision that directors must have a considerable shareholding. A valid criticism of the present position of joint stock companies is that the risks have been greatly reduced by amalgamations, Government support or the achievement of a near monopoly in their particular line. The managers feel only the effects on their prestige if their company is unsuccessful and the shareholders are little more than rentiers. In such a case why not nationalize? Admittedly even in such a case there is usually no reason to suppose that nationalization will be an improvement, but the free enterprise system is meant to offer some rather more positive inducements to efficiency than prestige. It is a system with checks and balances. If the system is to work there must be responsibility. The managers and owners should directly feel the consequences of success or failure. If the managers were to become too like Civil Servants and the shareholders almost negligible the system would work imperfectly.

But luckily this system is not part of the natural law handed down from heaven. It is a man-made creation, dependent on that most useful legal fiction, the joint stock company. Today it is time that we adjusted the company laws to give the workers some status as well as the shareholders. And Liberals want to see the growth of a body of worker-shareholders taking part in the affairs of the company in both capacities. In this way property can be more fairly spread, unity of purpose achieved and responsibility made real.

The full effects of such a change in industrial relations would mean that the workers came to regard themselves as employers of management. The remuneration of directors would be as subject to negotiation as wages. Profits would resume their true role as the residual reward for saving and risk-bearing. A confusing feature in disputes about wages would be removed. At present it is a favourite argument that higher profits generate a demand for higher wages. But though psychologically this may be true, it is not right to treat profits and wages on an equal footing: for profits are not a cost. Indeed the whole method of wage negotiation would be greatly eased. The wage-earner would have an interest in the free enterprise system and an understanding of his firm's position. There is, of course, no intention to depress wages by giving a share in the profits. On the contrary, as high wages should be paid as can be justified. But instead of wage negotiations on a national level it would be easier to deal with them on a more manageable basis. It would also be easier to relate them to the productivity of the business. While national minima should be established, Liberals see no reason why over the minimum wages should not be agreed on a narrower front than at present. The negotiations to fix a wage for a vast group of workers engaged in all sorts of jobs in firms of varying productivity leads to much friction and unfairness. Why should a man working in an engineering shop in Suffolk, just because he is classified as an engineer, be tied to the same wage as a worker in a shipyard on the Clyde? It is true that many workers get more than the agreed wage, but the absurdity of the system remains.

In their proposals for spreading ownership Liberals aim both to give everyone a share in the capital wealth of the country and also to give workers in a particular firm a share in the capital of that firm. And their proposals are directed to doing this by tax incentive, alteration of the company laws and various other measures designed to make it easier to buy and sell suitable securities.

The tax incentives must obviously be varied in detail according to the taxation rates and the taxes in force. As things are at present the first stage would be to remove some of the handicaps which deter firms from making shares available to the workers: these exist both under Schedules 'D' and 'E' and under Profits Tax. While a cash bonus, which achieves none of the advantages of ownership, is cheap to employers (for it can be charged to expenses), the distribution of shares at a reduced price or no price may attract extra taxation.

The worker, too, is apt to be mulcted in tax as his shares will be treated as part of his remuneration: though it should be said that

CO-OWNERSHIP

in some cases agreement has been reached with the Inland Revenue whereby the company pays a lump sum in settlement of the tax before the distribution of shares.

The positive incentives which the Liberal Party has suggested would allow companies to set against their profits for taxation purposes 110 per cent of any contribution to a scheme of employee shareholding. The types of scheme for this purpose are numerous: under some the employees receive their shares direct; under others shares are acquired for them by trustees. Sometimes the company adds to or subsidizes savings made by employees. An excellent summary of the principal categories of such schemes is to be found in *The Challenge of Employee Shareholding* by George Copeman (Business Publications, Batsford). It is true that with sufficient ingenuity firms can often get round the tax obstacles, but Liberals feel that we should go further today, remove the need for such ingenuity and the hazards it entails, and give recognition to the desirability of such co-ownership.

On the workers' side the Liberal Party has adopted a scheme of 'Employees' Savings Accounts' devised from Dr. Copeman's ingenious suggestions for a 'save as you earn' plan. Under this plan an employer pays part of his profits into a blocked account in his employee's name. The employee is not taxed. He may instruct the bank to invest this money as he likes. He may buy shares in the company in which he is employed, or in any other company or in a unit trust: he may prefer to have the money on account or invest it in bonds or Government securities. Only if he withdraws the money will he be liable to tax. This plan is an incentive to saving. It enables the worker in a small business, or in a State business, the professional man or the office worker or the self-employed to gain a stake in industry on favourable terms. By this scheme, too, the expense of buying small packets of shares can be much reduced. For the banks will, in aggregate, have considerable orders though made up of small individual demands.

The expense of handling small orders for shares is one of the main deterrents to an increase in general shareholding. Liberals have therefore proposed the abolition of Stamp Duty at least on small transactions. This duty is a relic of the days when shareholding was considered appropriate only for the rich.

CO-OWNERSHIP

Having encouraged the spread of ownership it is also essential to recognize the changed status of the worker by appropriate amendment of the company laws.

The other steps designed to make it easier to buy and sell shares in industry are such reforms as the freeing of the Stock Exchange from some of its inhibitions against advertising. Subject to suitable safeguard against the peddling of bogus shares there seems no reason why the advantages of investment in trusts should not be much better known to the public. Indeed the trust, whether unit or managed, seems an ideal investment for a man with a small capital which he does not want to commit entirely to one basket. Buying and selling in some trusts is already done across bank counters and might be extended to post offices. There is also a case for giving encouragement, by tax concession, to large companies which can themselves set up investment trusts for their workers.

But the process is slowed down both by the taxation penalties and also by the general ignorance of the free enterprise system. Although the changes in our taxation would not have to be great in themselves they could turn the steady trickle of co-ownership into a flood, especially if accompanied by a campaign to publicize the advantages of shareholding: a campaign in which it is to be hoped the Trade Unions would play a prominent part. The following extract from a letter by a shop steward to the *Manchester Guardian* shows both the lack of and the thirst for knowledge which exist.

'I am a regular reader of your newspaper, and as an A.E.U. convener I often read useful information in the Finance and Industry section. From time to time I am left wishing I had more information on some items, and I am wondering if it is in order to write for advice or information.

'For instance, enclosed are details of an offer to purchase shares by our firm. Frankly, most of it is above my head, and I am unable to answer questions such as "How do I sell my shares?" "Does the purchase of shares entitle me to vote at shareholders' meetings?" "What rate of income tax is paid on the shares, and/or the dividend?", etc.

'At our factory we receive a production bonus of £1 17s. 6d. per week. If employees could afford to put this money in the bank or

CO-OWNERSHIP

post office, would it earn less than shares in the company (if shares were given in place of bonus)?

'Superannuation and profit-sharing schemes are increasing and I have increased my knowledge of such schemes through reading your paper. Unfortunately, trade unions are not educating their members on these important subjects. Their advice is usually, "We can see nothing really wrong in these schemes. It is purely a matter for the individual to decide." Workers, however, expect their representatives to be able to answer every question and suspicion. Yours, etc., SHOP STEWARD.'

There were, in 1956, equity shares to the value of about £9,500 million quoted on the London Stock Exchange. It has been estimated that less than 150,000 persons among them own half these shares: while, of course, many more are owned by pension funds and trust companies. An extra 10 million people in the country owning £500 worth of equity shares each would absorb over half the available stock. But it is to the capitalization of reserves and the expansion of industry that we must look for the bulk of the new shares. The incentive given by co-ownership itself and the elimination of bad relations in industry to which it will contribute, should make our expansion more rapid and certain. Nor will the demand for equities for insurance funds continue to rise indefinitely.

The transition from an industrial oligarchy to an industrial democracy will be easiest made in a time of expansion. While Liberals want a more equitable sharing of the existing capital wealth of the country, it is to the new wealth in expanding trades and industries that they look to provide the bulk of industrial property for the new property owners.

One advantage of the Liberal scheme is that it will give the incentive to expansion, and as the expansion takes place the workers will reap the reward. We hear again and again the plaintive cry, 'How are we to achieve expansion without inflation?' There is a way. As I have said, profits are not a cost. The distribution of extra purchasing power through profits does not therefore contribute to cost-inflation. For in a competitive society profits will only be earned by producing something the consumer wants at a price he is prepared to pay. The extra goods, for which there is a ready de-

CO-OWNERSHIP

mand, must be purchased therefore before the profits arise. So long as we match any extra income by extra goods to be bought, inflation will not arise. Again, profits are a direct incentive to harder or more efficient work when they are the profits of the workers' own business. So while expansion is essential to Liberal plans, it will also be a result of these plans.

It will be the workers in large public companies who, in the first stage at least, will become shareholders in the firms for which they work. We do not pretend that it will be possible in the near future to make every farm servant, every employee in a small private company, a part-owner. The initial plan will be to spur on the movement already seen in such businesses as the John Lewis Partnership and I.C.I. But there is no reason why small firms, farms, shops and all manner of miscellaneous businesses should be excluded. For one thing many of these businesses are limited liability companies, subject to Profits Tax and therefore susceptible to Liberal incentives. Workers in every sort of occupation will be entitled to take advantage of the 'Save-as-you-earn' plan. The industrial workers in the bigger companies will by no means be the only people to gain from such a plan. Further, as the climate of ownership permeates society, as capital gains spread in the hands of the many, it will be easier and more natural for all sorts and conditions of men to own a share in all kinds of business.

The corollary to ownership is, as I have stressed, responsibility. The recognition of the worker-owners' status will accentuate the trend towards joint consultation at all levels. The trend will be magnified out of all recognition. For joint consultation will no longer be between 'the two sides'; it will no longer be a conciliatory gesture by the good employer, or a convenience. It will be natural, inevitable, the discussion between partners, as to how they are to conduct their partnership.

I must now deal with some of the objections to the Liberal Plan.

There will be losses as well as profits. The new owners will sometimes find their shares fall in value; that is agreed. But let no one think that Socialism will guarantee them against losses. If the State buys shares in industry, or undertakes direct nationalization it, too, risks losses. These losses will fall on the taxpayer, which is to say the public at large. The present system by which equity ownership is concentrated in very few hands does admittedly concentrate the risk of loss on a small class. This is, however, no reason for perpetuating the present system. For one thing, taken over the years, equity investment has proved rewarding. Nor is investment in equity shares the only way of losing money. Buyers of government stock have lost. Wage-earners, of course, lose when the firms for which they work go bankrupt. Householders lose if house values fall. All lenders or holders of money, or those who draw fixed money wages, lose when the Government inflates the currency. It is beyond human ingenuity to prevent all possible losses. Liberals certainly would claim neither that it can be done nor indeed that it is a particularly important objective to try to reach. The important thing is to give a fair chance of success; to see, as far as possible, that those who deserve success, get it, and to look after any who are unlucky. But losses in their property or on their profits will not reduce the new property owners to penury. Those in employment will still have their wages. Those in self-employment will be no worse off than they are now. Nor will the worker-owners be compelled to take shares in their own or any other business. The facts remain that to run a free system you must interest more people in it; to achieve efficiency in industry and stability in society you must promote industrial partnership; and finally the 'have-nots' want property. The terrors of ownership have seldom deterred people from accepting it. Nevertheless, we must give workers a chance to own shares in other businesses besides that in which they work. The bankruptcy of a business in which a man not only worked but kept all his savings could be serious for him. As I have shown, Liberals are anxious to see more property of all kinds in more hands and as great a spread of savings as possible.

Another objection raised against Liberal schemes is that they will not make much difference. Even if a man has a house, 500 or 1,000 pounds invested and a considerable array of 'consumer durables', a television set, a washing machine, a car and good furniture, yet it could be said that all this will not add up to much when set against his weekly wage. Indeed, as the house may be mortgaged and the television on hire purchase, modern practice

CO-OWNERSHIP

may actually increase his dependence on the weekly wage. It is said, for instance, that it is the regular pay cheque which is the great bait to the farmer to produce milk.

This brings us up against a perennial difficulty of reformers. Their reforms are apt to be judged against an unreformed background. Of course, if everything else remains the same and co-ownership is simply added to the existing setup, its full effect will not be felt. But everything cannot remain the same. It is like postulating that if I become a strict Muslim, I shall find it very difficult to continue to drink alcohol. If I became a Muslim I should presumably alter my habits in a great many ways and I should have reconciled myself to teetotalism. The atmosphere of industry and of everyday life will have changed. It will be normal to own a substantial amount of property. It will be inevitable that people will adjust their outlook. The wage or salary will still be immensely important. But do not let us underestimate the margins of life. A comparatively small amount of savings can be immensely important to a man, though his earnings may be high. Liberals see nothing inconsistent between property ownership and enhanced earnings. Indeed a paradox of our present situation is that while the power or responsibility attached to property has diminished, the advantages of having any capital are as great or greater than they have ever been. This is due to the high rate of taxation and the heavy taxes on earnings. I do not want to discuss this here. But it is relevant to point out that if you make saving profitable you reduce the need for some of the social services which are at present provided out of taxation. The possibility of acquiring property depends on reducing taxation, and the wider distribution of property will mean that very high marginal rates of taxation are unnecessary. The growth of hiring and hire purchase in all its forms is an important economic factor whether co-ownership grows or not. It is no new thing. But it is now of very considerable importance in our economic life. On the whole it is welcome. It enables people to live at a higher standard of life than they would achieve without it. That this is due to a persistent deception, so to speak, does not detract from its merit. When it comes to buying furniture and small household goods on hire purchase, the buyer would very often be better off if he banked the equivalent of the weekly payment and

CO-OWNERSHIP

only bought the goods when he had the cash. The trouble is that he might not be strong-willed enough to save the money in this way. Hire-purchase, like life insurance, can be a stiffener to the savings instinct and as such is good. Where it becomes dangerous is when a man overreaches himself, or when it snowballs. The 'never-never' which is almost literally never, never, because the contract is one of hiring, or because the goods are constantly being replaced by bigger, better or just shinier products can be a menace. It can reduce a man to a slave of his appliances. His weekly wage is eaten up. He dare not break the chains because the penalty is severe. He is reduced to a state of dependence and constant anxiety. Liberals see no objection in principle to imposing regulations on the hire-purchase system. But the regulations should be against its abuse. In essence, it is a reasonable economic facility and compatible with the wider distribution of property.

Another line of objection attacks Liberal proposals from just the opposite direction. Over against those who say that ownership and responsibility can never be more than marginal for most people, stand others who say that the proposals would be too revolutionary. In the mind of the critics is often a vague feeling that because many units of industry are big, they must be owned by something big. It is as though you expected only a few big men to live in a big block of flats and not a lot of normal-sized men. This is obviously in its simplest form a hallucination. I.C.I. already has 260,000 shareholders. It could get on with five times the number. Indeed, co-ownership is easiest to work where the unit is very large. But, of course, if we reached a position in ten or twenty years where every family was a substantial owner of property in general and of a decent share in industry in particular, it would be a revolution. It would trench on the preserves of many vested interests. It would reduce dependence on the State. It would alter patterns of demand. Belloc, when advocating his earlier form of co-ownership through Distributism, thought that the vested interests would find the prospect so alarming that they would pull all the strings open to them to frustrate it. As he was convinced that the strings to be pulled in an industrial democracy were strong and numerous, he sometimes wrote as though he had reached the conclusion that Distributism was doomed to failure.

CO-OWNERSHIP

In fact, however, the conclusion to be drawn from his writings is that Distributism would radically alter the particular type of Democracy he knew—and alter it for the better. The antipathy which some Trade Union leaders sometimes show to co-ownership is easily explicable. Ever since Marx, nothing has been more alarming to Socialists than the possibility of the disappearance of the proletariat. Nothing draws sharper cries of anguish from the Marxist than the fear that the masses would acquire middle-class values. But co-ownership would only threaten the faults of Trade Unionism and not its virtues. It might well decrease the automatic acceptance of allegiance to Socialist parties. It would break the more restrictive practices of the Unions, for the workers would see that these were against their own interests. It would undermine the trouble-maker and the so-called leader who can think of industry only in terms of class warfare or the maintenance of a Union closed shop. But all this would be to the good. On the other side of the penny the Unions would retain their useful functions in negotiations over wages and conditions. They would be encouraged in their new role of collaboration with management. They would, I should hope, remain not only watchdogs against abuse of industrial power but also goads to greater efficiency in management. As for the effects on the role of the State, the revolution would be wholly beneficial. And there could be no better way of breaking down the worst features of class stratification.

There remains, however, a deep-rooted suspicion of Syndicalism. I have never understood this. 'Syndicalist' is a word that frightens British Socialists as much as British Conservatives. When they set up the Nationalized Industries, the Labour Government were particularly careful to forbid all possibility of Syndicalism. They abolished profit sharing in the gas industry. They created nationalized boards on the pattern of the directorates of private industry. They maintained all the features of the divided structure of industry—the workers and the bosses. They did this for various reasons—innate conservatism; a hangover from the Marxist heritage; but also from a fear that if the workers in a particular industry came to dominate it the general welfare of the people at large would suffer. This last reason deserves consideration. In Socialist conditions there is a great deal to be said against

CO-OWNERSHIP

Syndicalism. The neo-Marxist Socialist believes in centralized control and planning to which Syndicalism may offer very effective opposition. The neo-Marxist also wants a closed monopolistic type of economy. In conditions of monopoly, workers' control could be dangerous to the public weal. But though personally I am not frightened of Syndicalism the Liberal proposals are not syndicalist. We do not anticipate the abolition of shareholders, not even the abolition of outside shareholders. We want most of the workers in a firm to join the shareholders in becoming joint owners. We do not suffer from the delusion that the workers can, themselves, direct the business. Though we are anxious to see management more closely identified with ownership, we do not say they are the same. We regard it as most important that, while the workers should be consulted at various levels, and while as part-owners they should be in a position to exert, in extreme cases, the ultimate sanction of dismissal, they should not normally interfere with the day-to-day running of the business. Liberals deplore the recent decay of managerial responsibility. They want management to manage. But equally we want co-ownership associated with a Liberal society and competition. Competition is the safeguard of the public. The Socialist attempt to protect the public interest by making the Nationalized Industries in some nebulous way responsible to Parliament, has failed. Competition has no equal as a protection for the public economic good, subject to a sensible general structure of industry. If competition is maintained the public need have no fear of one industry or firm establishing undue power and holding the people at large up to ransom.

The objection that the workers will not respond is not backed by evidence. The comparative buoyancy of savings against the dreadnoughts of inflation and high taxation is a sign that with any encouragement people in a free society will save, and save substantially.

There is no more harmful propaganda than defeatism about savings in a democracy. The notion that we must surrender to an autocracy so that the Government can force us to save, is pernicious. Nor is it true that the workers are not interested in becoming shareholders. In fact, 75 per cent of the worker-shareholders in I.C.I. retain their holdings. And this is achieved in a society which

CO-OWNERSHIP

is still coloured by prejudice and ignorance so far as equity shareholding is concerned. Given proper education and information about a fair, free enterprise system, it is only those with a singularly low opinion of their fellow-men who think that they will invest in the pools but not in industry.

It may be more difficult to get all workers to take responsibility. Responsibility has been undermined in every direction. But here, again, experience is encouraging. In the John Lewis Partnership the bulk of workers take effective interest in how the business is going. Consultation is spreading rapidly in industry; as education grows its spread will become more rapid. The Liberal proposals, once launched, will find a tide in their favour.

This brings us to those who say that if it is advantageous for men to have a bigger share in the ownership of industry why must they have special incentives? Out of good wages those who want shares can buy them. Why should you force the taxpayers at large to shoulder an extra burden so that people who have not enough sense to save and invest of their own accord may be induced to do so? This would be a powerful argument if it were the case that you could ever hold the scales impartially in this context. It rather assumes that the free enterprise system can be run in all its pure theoretical beauty, and that in this idyllic state it is the natural form of economic life. But I must reiterate that the free enterprise system as practised is artificial. It is subject to influence by taxation, social habit and the general law. For many, many years, the scales have been tilted against private saving and the spread of ownership. There are plenty of examples of how this has been done. There has been the prevailing current of Socialism. There has been the high taxation on earnings. It is almost impossible to-day for anyone to accumulate capital by saving out of earnings. In spite of the prejudice against capitalism, never has it been so great an advantage to own some capital. In spite of the vague but strong feeling that honest toil should be rewarded and speculation discouraged, a win on the pools is, financially, the best thing that can happen to anyone to-day. Capital is built up by inheritance, by any sort of capital gain which goes free of tax, and by the self-employed. These methods are not open to many of us. The result of Socialism has been to confine the acquisition of capital to the

CO-OWNERSHIP

lucky, the gambler and those who have special skills, sometimes in exploiting an unfilled want but often in simple financial manipulation. This is no coincidence, nor will it be cured by more Socialism —unless, of course, we are to be taken all the way to Communism —it is the result of the present ambivalent attitude towards free enterprise; the attitude of fond disapproval combined with a determination to subject it to high taxation in an all-pervading State. Financial policy, too, has favoured the big established company. The credit squeezes have not affected the company with big reserves. Company taxation has favoured the ploughing back of profits rather than their distribution. For many years it has suited the Government departments to deal with large rather than small firms (I have experience of this in my own constituency). As the Government is a main source of orders, their general preference for the large order and the large contractor has contributed considerably to the growth of 'oligopoly' in British industry. At every financial crisis, under Tory or Labour Governments equally, the screw has been turned on house ownership and hire-purchase, two of the most popular roads to property ownership. The nationalization of some industries has contributed to the concentration of economic power.

The free enterprise system has never been and can never be run free of all interference. It must have a framework of law. Lately that framework has been inimical to the spread of economic power and ownership. There is no reason to suppose that this is due to anything except to lack of understanding of the system coupled with indifference and a climate of Socialism. If we want ownership spread, positive steps must be taken to achieve the spread.

But certainly the aims and limitations of co-ownership must be kept in mind. I notice in some writings on the subject the authors slip into mystical language worthy of Plato or Hegel. For some of them 'the company' once it is transformed into a partnership on the 'John Lewis' or 'Carl Zeiss' model becomes a good thing in itself. For my part I regard any company as simply a convenient means to satisfy some demand. It should not be left in existence after the demand has been met. Today companies tend to grow personalities and personalities which are very strong. Their purpose is lost sight of. A big company tries to build up loyalties

CO-OWNERSHIP

which will sustain it. This is good in some ways. But it also leads to the demand that if it gets into difficulties it must be helped. It becomes a part of the British way of life; its success is a matter of prestige. The spiral is very effective. The more imposing the company the more its power, and the more powerful its influence the bigger it gets. Modern companies, too, try to attach their workers by supplying houses, sports grounds, medical service, pensions. This, too, is excellent so far as it meets a need and helps the worker. But in so far as it turns the company into a benevolent despot and the workers into its dependants, it is not good. It could have bad effects on the flexibility of our economic system. More important still to Liberals, it softens personal responsibility and restricts freedom. We are not saying that the man, cradled, nourished, entertained, housed and buried by a big company, is necessarily a proletarian slave. But we do say that the end of industry is to provide goods for the consumer and in doing so to provide income for the producers. Industry is not an end in itself. Nor is any industrial firm.

But it is not inevitable that if co-ownership is introduced it must lead to the provision of all sorts of services in kind instead of payments in cash; it need not increase the ossification of industry nor need it encourage the glorification of firms as ends in themselves. The chief safeguard against these possibilities should be competition. A further safeguard will be the spread of the workers' savings among several investments. There certainly would be disadvantages both economic and political if the workers in every firm became extremely identified with the firm. If they lived in the firm's houses, worked in it, relied on it for their pensions, and looked to it to fill in their leisure, the resistance to change might be very great while formidable political pressure would be built up if the firm showed signs of going into liquidation. But if the worker has a share in the equity of many companies, if he is the owner of his house and the provider of his own amusements and pension, there is no reason why such pressure should build up.

But what happens if the worker's company makes losses? Granted that losses are a risk we all run, will they not be peculiarly damaging under co-ownership? The implication of this question is that if the worker does not save and invest, he will not make losses.

CO-OWNERSHIP

This is nonsense. Somebody has to save and invest on behalf of us all. If we rely on forced saving thrust on us by the State for investment in its own projects, we may lose just the same, as I have said. We have all lost millions of our savings since the war; the Government has lost hundreds of millions of our money; examples of loss are the Ground Nuts Scheme, the railways and the aviation industry. Nor can the worker avoid loss by investment in gilt-edged securities—as we well know. But the further implication behind the question is that there is a class of people who understand and like risk-bearing and a class of people who cannot and should not undertake it. This division is a reflection of the division of industry and society, the class system which Liberals want to break up. Of course, if workers have low wages and very little property, they will not want to risk what they have. Of course, if you perpetuate a class of millionaire property owners it is best that they should take the risks. But we want to end all this. We do not pretend that everyone will take a continuous interest in stocks and shares. We do not pretend that anyone will like making losses. But we deny that the majority of the population are incapable of understanding economics and so will be panic-stricken by a fall in the value of a share; nor do we believe that insulation against loss is possible or desirable. The importance of our proposals in any event does not lie solely in the possibility of some extra gain, whether by dividend or capital gain. If they are regarded as a method of providing some Danegeld to buy off the critics of a fundamentally shaky system, they will fail. But it is the new status and responsibility, the ownership aspect of this which is so important. The dismay which may be caused by losses has been exaggerated. The overwhelming advantage of having some share in ownership and profits must be apparent to anyone who studies the movements of equities as against fixed interest securities for the last twenty years.

The whole case for co-ownership depends on a belief in the free enterprise system. If co-ownership is to be an attempt to destroy the system it will fail. Some confusion exists on this point. Belloc and some early distributists wanted a wide distribution of property, not only to avoid slavery under the Servile State, but also as an attempt to return to some more primitive form of society. To-

CO-OWNERSHIP

day there are advocates of co-ownership who seem to hanker after the bliss of the same sort of archaism. There may be a lot to be said for crafts, for less commercialism, for a society in which status has not given way entirely to contract. In this chapter, however, I am not concerned with these various possibilities. Co-ownership will not, in itself, relieve the boredom of work on a conveyor belt. It is compatible with production by conveyor belt. It is not a retreat from the opportunities with the necessary uncertainties of the modern world into some more supposedly sheltered past. It is not an attempt to reverse the division of labour. Nor is it founded on any view that the private system has gone wrong and become in some way immoral. I can see nothing immoral in a man who has the means to venture on some business agreeing with other men to work for him for a stipulated wage and for still others to lend him money in return for a dividend if profits are made. Apart from the full-bodied attack of Marx on free enterprise, there are critics who argue that as shareholders take little part in controlling the business they should not be treated as owners, and since workers contribute very largely to its production they should be. Liberals certainly want to amend the company laws, and they certainly want workers to be egged on to become part-owners. But it seems to me that there is a confusion of thought when this argument is pressed to the point where it is suggested that shareholders are immoral and should be virtually expropriated. This is a return to status with a vengeance. As I have said, why should not a man contract with others to lend him money on condition that he gives them some of the ensuing profit? The lenders certainly perform a service. If the contract is that they be associated as owners—why not? The trouble does not lie in the morality of absentee ownership. The trouble lies in the inability of most people to become owners whether absentee or not. When Liberals want to give workers status, they want to do this because they contribute work; in so far as they become owners they will have rights under that head, too: but Liberal reforms are not based on some perverted version of the Marxist theory of value by which workers, as workers, are treated as owners. There may well be entrepreneurs who do not want to share their profits and rights of ownership, whether with workers or absentee shareholders. They

CO-OWNERSHIP

will be free to carry on, though they can hardly complain if the full protection of limited liability is not granted to them.

Co-ownership, with its save-as-you-earn plan, will be an incentive to efficiency and an incentive to saving. It involves a tax on expenditure of a very much simpler and more acceptable type than any other yet devised. It does not involve a continual gamble on the Stock Exchange. The free system, like any system, depends on carrots and whips; its distinction is that the carrots and whips are more humane, as well as being more effective than any others yet devised. Admittedly it demands of human beings that they should behave rationally and take a rational interest in political economy and their own personal finances. Some people may think that this is asking too much. They will point to dead-end kids all too ready to take dead-end jobs and to workers who are only anxious to be freed of any responsibility. But these critics overlook the enthusiasm of the great majority of the new generations, their inquisitiveness and their self-reliance. Already within industry there is growing up an atmosphere of co-operation far removed from the class war.

6

The Social Services

Our Social Services were the invention of a Liberal Government. They were brought in as a piece of social engineering to meet an obvious need. Liberals in office in the early years of this century saw that poverty was serious and widespread. They set about relieving at least some of it to a small extent. Mr. Asquith said in the House of Commons in 1906:

'To my mind the two most tragic things in our modern social condition are the figures of the man who wants work and cannot find it and of the man who is past work and has to beg for his bread and his bed. So long as these figures remain in the foreground of our life here in Great Britain, they constitute a standing reproach to our civilization.

After steps had been taken to strengthen the financial position Mr. Asquith found it possible to provide for Old Age Pensions in the 1908 Budget. Several countries, including New Zealand, Germany and Denmark, already had State pension schemes. These schemes and the problem at large had been studied for many years before Mr. Asquith.

The genesis of the Social Services was an attack on poverty. They were not originally conceived as a method of redistributing income, nor related to any philosophy of the State other than the general programme of political and Social Reform to which the Governments of Sir Henry Campbell Bannerman and Mr. Asquith were pledged. When introducing the pension scheme Mr. Asquith said that, 'any practical scheme must be based on a certain discrimination'. It is also interesting to recall that he told the House of Commons, 'All so-called contributory schemes must be ruled out.' His reasons for both these decisions were practical.

THE SOCIAL SERVICES

The Liberal attitude to the Social Services has continued very much on the original lines of 1906. We regard them as empirical steps taken to meet particular situations. We believe that their needs are a sufficient justification for helping the unfortunate. It is the unfortunate, the poor, the sick, the injured and the unemployed whom we have wanted to help. We have been prepared to extend the services offered by the State to other classes where the need is apparent and is not being otherwise met. The Liberal Party espoused the cause of family allowances when the Labour Party would not do so because of opposition within the Unions.

Of course, the effect of Social Services has been to bring about greater economic equality. Liberals seeing the gross disparity of wealth which existed in Edwardian times, and indeed much later, were pleased that this should be so. They fought those Tories who opposed the Old Age Pensions, Unemployment Pay and Health Insurance, and in the course of the fight privilege and wealth were denounced. They believe that a degree of economic homogeneity is essential to democracy. But Liberalism is not deeply concerned with precise equality of incomes. In so far as it believes that too great discrepancies in wealth distort society, it wants to achieve greater equality by the redistribution of property rather than earned income. Liberals certainly believe that taxation should be to some extent progressive: and that a secondary reason for this may be to achieve a fairer distribution of purchasing power. But their main concern in this field is over what they regard as the unfortunate premium which now attaches to luck in gambling or inheriting wealth. The Socialist sees the Social Services as an inherent aim of the State. Bismarck, who may claim to have started them in their modern form, viewed them in this way. As well as relieving poverty they have further advantages in the eyes of the Socialist. As well as being a means of redistributing purchasing power, they enable the State to assert the equality of citizens before it. They can be used to break up class distinctions. The Socialist, too, feels no qualms about using them to influence people to do things which are for their own benefit, though they would not do them if left to themselves.

These additional advantages, as they seem in Socialist eyes, are not negligible. They have had their effect on Liberal thinking.

THE SOCIAL SERVICES

There is a lot to be said against the old treatment of paupers as an inferior class disgraced before their fellow-citizens. No one, Liberal or not, wants to go back to the days of the old Poor Laws and the Workhouse. There always was and still is a case for making unemployment insurance compulsory to all, and family allowances universal. There is a practical case—State insurance is difficult to run on a voluntary basis. But there is also a theoretical case. We want to fuse the nation together rather than to encourage class divisions and 'state-aided'—'non-state-aided' is such a division. Nevertheless, while these arguments are strong, the arguments against encouraging the reduction of citizens to proletarians are stronger. While there may well be exceptions, Liberals must stick to the general rule that social services are primarily designed to fill some gap, to meet some need, and are not good-things-in-themselves to be extended far beyond the point at which the need is met.

The use of the Social Services to influence or compel people to do things which they might not be strong-willed enough to do for themselves can also be supported up to a point by valid reasoning. The catch phrase about the 'Gentlemen in Whitehall knowing best' has been rather unfairly thrown in the teeth of the Labour Party. In its extreme form it typifies an absurd attitude which is common among planners, but it also can have an element of truth in it. The food subsidies and the provision of meals at schools indirectly interfered with the individual's right to spend his money as he likes, but they have done a lot to improve health. Messrs. Rowntree and Laver estimate that between 1936 and 1950 the proportion of the working-class population of York who lived in poverty fell from 31·1 per cent to 2·77 per cent; if increases had not been made in welfare legislation it would have fallen (through higher wages, etc.) to 22·18 per cent. Even allowing for the margin of error inherent in these figures it shows that welfare legislation achieved a large measure of good; such legislation included Children's Allowances and food subsidies. These welfare schemes were not merely additions to income; the Gentlemen in Whitehall decided that some things needed encouragement and to some extent these gentlemen were right. But again while not being doctrinaire on the matter Liberals must be inclined to favour the type of assistance

THE SOCIAL SERVICES

which contains the minimum of 'truck': as a general rule subsidies are best paid in cash. If the recipients make bad use of the cash we should look first at the causes of their folly before deciding that they should be relieved of the responsibility for making their own decisions.

To a Liberal, then, it is perfectly proper for the State to intervene to right an obvious wrong. Such efforts by the State should be directed to enabling the people concerned to surmount their difficulty and the need to encourage the individual to assist him to choose a better way of life should have priority over anything else. Just as a humane and Liberal approach does not primarily flow from a desire to extend the scope of State action, or to force people into some pattern of expenditure near to a Platonic ideal laid up in some planner's White Paper, so Liberals should free themselves of any slavish attachment to *laissez-faire* in this connection. Unemployment Insurance is no doubt an interference with the free play of the labour market. So much the better in my view. While 'the greatest happiness of the greatest number' is a useful general motto, and particularly valuable when it is suggested that a certain class should have their already considerable happiness enlarged at the general expense, I reject it utterly as a reason for tolerating acute unhappiness when it could be cured by a very slight decrease in the happiness of the majority. In economics, as in politics, minorities have their rights and it is peculiarly the business of Liberals to look after them.

Now let us see where we stand today. There are many activities of the State which have an element of Social Service about them. I rule out for present purposes subsidies paid largely for prestige reasons to certain industries such as film production. But in nearly all the nationalized industries there is a flavour of social service. These industries borrow cheap, the Government ultimately stands behind any losses and in several cases prices are held down on the rather unformulated ground that it would be 'unfair' to charge them at the full rate. The railways and the coal industry are expected to provide some services cheaper than the market might allow so as to benefit their customers. Then there are the subsidies to particular essential services: there is a subsidy of about £350,000 a year to McBrayne's steamers in the Western Isles: there

THE SOCIAL SERVICES

are the Housing Subsidies: and there is Education. These are not usually counted as Social Services (nor are Agricultural Subsidies), but their justification is much the same as that for a particular social service, that is, that some ill would exist, harmful to a section of our people, if they were removed. The Social Services proper are usually considered to be Unemployment Benefit, Sickness Benefit, Pensions, Maternity Benefit, Widows' Pensions, Guardians and Death Grants, the National Health Service and National Assistance.

The income of the National Insurance Funds (less payments for National Health) were in 1956–7 about £693 millions. The funds were slightly in surplus, outgoings being about £658 millions. But this is far from being the whole story. Apart from those payments mentioned above which have an element of Social Service in them, the Health Service costs about £612 millions and National Assistance £134 millions. Of the annual budget, therefore, something like 33 per cent (or 13 per cent of the national income) goes on Social Services, even if we define these fairly narrowly. We must now ask what is the logic behind the taking of so high a proportion of everyone's earnings for expenditure on Social Services. 'In a liberal society the basis of any policy regarding social policy is the belief that all persons should live in reasonable comfort within the limits imposed by the state of development of the economy, and that no person's opportunities to develop his particular gifts should be frustrated by material circumstances.' So writes Professor Alan Peacock who has made as close a study of the Welfare Services as anybody since Lord Beveridge, and we may take this as a good general rule. Professor Peacock goes on to argue that the liberal attitude to the Social Services rests on a positive dislike of gross inequality and that, 'A liberal social policy designed to maintain a minimum standard of comfort may require us to intervene positively in the system of public licence in order to redistribute income, but it justifies no other form of social services other than those which involve tranfers of income.' I agree with Professor Peacock on his fundamental point, that Social Services should increase the income of the poor where those incomes are inadequate to support a reasonably comfortable life. And as I have said, this involves a move towards equality—which, in itself, is no bad

THE SOCIAL SERVICES

thing. I also agree with Professor Peacock that gross inequality is illiberal. But I would suggest that the word 'gross' begs the question. And I would maintain that today it is still the Liberal view that the purpose of Social Services is not primarily to take purchasing power from the rich. If this is the fundamental Liberal attitude towards the Social Services, there is nothing sacrosanct about their present form. Further, those which are directed to a specific object must have some special justification. Finally, if and when we are all endowed with opportunities to become rich enough the general services can wither away leaving only provision for those who cannot take or have not taken their chances.

Looking to the very long run Liberal principles give a definite shape to our thinking about these services. We should not assume that they will all always be necessary. As real wages rise people will be able to provide to a greater and greater extent for their own pensions. Already more and more people are being taken into voluntary schemes. Liberals should beware of schemes which presuppose the indefinite existence, not only of subsistence pensions, but of pensions at a considerably higher level. I am myself sceptical of the advantages of funding State pensions, partly because it means that you will indefinitely perpetuate a system which may, in the course of time, be neither necessary nor Liberal; partly because I am sceptical of the whole funding argument. In real terms, of course, there is no saving in the sense of savings which can at some future date be used. The importance of saving is that it withdraws some current demand. In a private pension scheme the bargain in crude terms is that I surrender some of my current demand, my current claim on production, in return for the insurance company surrendering some of its claims to me at a later date. It seems to me at least highly arguable that in due course it will be more realistic and convenient to pay such State pensions as are necessary out of taxation, leaving it to individuals to enter into private bargains with insurance companies or their equivalent as may suit them.

As regards health the same argument applies to some extent. In the long run we hope that people will have enough money to pay their own doctor. Are there any special reasons why, when this happy state is reached, they should not do so?

THE SOCIAL SERVICES

Yes there are at least three arguments in favour of maintaining a National Health Service, even if we were all as rich as rich could be. The first is that this is a field where some gentle compulsion ought to be brought to bear on the individual. Children should not suffer, for instance, because parents are parsimonious, and even the very rich can be very parsimonious. Secondly, the community has a great interest in preventive medicine, medical research, etc. Thirdly, owing to restricted demand leading to a 'natural monopoly' in some areas, and because of the nature of medical practice, it is more efficient to treat it as a public service. In a limited field these and similar arguments may be conclusive even in the very long run. Specialist services, research, hospitals, the general prevention of disease and the promotion of good health will probably always be best organized publicly. I say publicly, though I would hope that they might be removed from the general stream of political argument. But again it may well prove that these services can be met entirely by taxes, subject to the general grading of taxation in general, rather than by the poll-tax of an insurance stamp. Already the proportion of the National Health Service paid for by the stamp is small. Ordinary everyday medical attendance, however, may well be a reasonable charge on a personal or family budget.

Very much the same considerations which apply to health can be applied to education. It has been suggested by Professor Milton Friedman and others that the main virtues of a national education system could be promoted by a system of standards and payments. That is to say, parents would be required to see that their children were educated up to a certain standard, when they would be eligible for a payment per child. But even some such scheme as this would involve the maintenance of a considerable staff of educational experts. The case for a complete system of State education seems to be strong, and likely to prevail far into the future. Equally, however, there seems no reason acceptable to a Liberal why State education should be the only form of education, nor why those who choose to have their children taught outside the State schemes should not be granted some assistance. As I deal with educational policy in general elsewhere that is all I intend to say about it now.

THE SOCIAL SERVICES

As I have said, the National Insurance Fund is now showing a surplus; soon, however, this will be turned into a heavy deficit. Even if there is no large-scale unemployment it is estimated that by 1979–80 the annual deficit, unless income is raised, will be running at about £475 million. This, of course, means that in all likelihood the actuarial principle will be increasingly ignored. This principle was behind Lord Beveridge's original proposals. It has been calculated that a pensioner who retires this year may receive with his wife some £3,500 in retirement pensions against a contribution of £85. To reinstate the actuarial principle will be politically difficult and economically it will put a heavy strain on individual incomes. The position will be aggravated by the latest proposals of the Labour Party. As far as pensions are concerned the Liberal solution is to aim at encouraging voluntary pensions, always allowing for a basic State pension in cases of need. This must, however, depend upon voluntary pensions being made transferable.

But as far as unemployment is concerned the case is different. Again, looking at the long run, Liberals look to a society in which our financial techniques will have lessened the chances of widespread or prolonged unemployment, and in which everyone will be enabled to save something against temporary loss of earnings. Up to now unemployment payments have been lower than was anticipated. Though we have probably been rather too optimistic about our success in keeping the economy for ever at full employment, nevertheless a mass failure of purchasing power and catastrophic unemployment does not seem now a probable calamity. What we have to provide against is unemployment in a particular industry or district due to changes in world supply and demand or improvements in technique.

The history of the housing subsidies is a painful lesson in economic muddle and political log-rolling. There may be a Socialist case, an extreme Socialist case, for treating housing as a public service. There is no need to rebut it here because it flows, if it flows at all, from general Socialist reasoning which Liberals reject. The more reasonable case for subsidizing housing rests on the assumption that owing to the shortage of land, etc., in suitable places its cost will be prohibitive to the lower-paid workers while good housing is something which is essential to the minimum of comfort to be expected in a civilized community. The stock Liberal answer, that you should build up the income of the would-be house-owner or tenant rather than put a subsidy on all publicly provided housing, is not entirely convincing by itself. For Liberals have always recognized the peculiar economic position of anything to which land space is important. Even if you subsidize the man who has not got enough money to pay the rent in a certain district he still may be unable to get a house. And who is to say what is a fair rate of subsidy? If my work is at Piccadilly Circus am I entitled to be helped up to a point where I can pay a Mayfair rent? Or am I to be sent to live at Uxbridge, and if so is my travelling to be assisted from State funds? As soon as you begin to consider the problem in these terms it ceases to be a welfare problem and becomes a matter of welfare plus planning. I believe that the State must intervene even more directly than it does now in certain planning problems and it is under this head of Town and Country Development that problems of rent, housing subsidies, slum clearance as well as some road and transport problems should be tackled.

This brief survey of the long-term development of the social services leads to the conclusions that eventually in a Liberal society they can be greatly reduced in scope and expense. With the exception of education, there is no compelling reason why they should not be concentrated largely on those who are in want of adequate income, for some cause or other. Unless we are to assume that large sections of mankind will never be fit to take responsibility for their own lives, there is indeed no reason why we should not look forward to the time when the only Social Service is a straight payment to bring everyone's income (including that of children) up to a national minimum. No doubt such a payment will always need to be accompanied by precautions against abuse. It may also be varied according to circumstances. Of course this will involve some investigation into a claimant's means as well as his willingness to work, etc. But a means test already exists for income-tax relief, educational grants and National Assistance. The means test of the thirties was objectionable because of the methods used, rather than because of any inherent indignity in the process itself. Admittedly it is often difficult to apply a test. A man

THE SOCIAL SERVICES

makes over his property to his son—should he be granted assistance from the State? Should a family be expected to use up all their savings before they get aid? Certainly there are difficult problems. They exist now. They will exist under any scheme of assistance. They are not a reason for showering benefits on everyone whether need is proved or not.

It is quite apparent, however, that while we should keep the ultimate pattern in mind we have nowhere near reached the situation where Social Services can be reduced to one subsidy to increase inadequate incomes. There is going to be for many years a need for increased assistance to the old. Even though as we get further from the thirties we may expect higher wages to be reflected in higher savings, we cannot in the immediate future expect the weekly wage-earner to be carried in old age by voluntary saving or insurance, even when backed by companies. We should allow for a subsistence pension tied to the cost of living. We should also encourage the growth of voluntary pension schemes by making it easier to transfer them when the man or woman concerned changes employment. We should also educate the public in the appreciation of the hard fact that the greatest enemy of the pensioner is inflation. On empirical grounds I see no reason why Liberals should support the retirement rule. There is a good deal of rather bogus mystique about pensions. Perhaps there is some logic in insisting on retirement as the pension is intended to cover the circumstances of retirement. But the logic of the whole insurance system is being thrown out of the window. The unfairness of the earnings rule is too widely felt for logic to weigh down the balance.

In the Health Service we should begin to introduce a sensible system of payment. On grounds of economy alone this is desirable. The present charge on the bottle, brought in after rather hurried consideration to meet difficulties of the Exchequer rather than the Health Service, is an unfair and barely defensible form of tax. Personally I think some payment by patient to doctor would be preferable. Liberals would like to see the relationship of patient and doctor strengthened. They do not like certain effects of the present system on the medical profession itself. Young doctors and surgeons find promotion is too much of a lottery. The system of appointment to vacant practices needs to be constantly watched. The scope for the doctor of genius seems too restricted. And the administration costs are very high. Nevertheless, on the whole a national system has much to be said for it, and will have much to be said for it for many years. Against any relief to the Exchequer from charges to the patient must be set the need for more and better hospitals, and preventive medicine. The emphasis ought to be placed on their side.

The recent improvements in our methods of dealing with unemployment are to be greatly welcomed. We should press on with all speed to deal with the man or woman who is faced with the need to move to a new district or to learn a new skill. It is time that it was possible to draw a considerable lump sum out of the National Insurance Fund, or at least to draw even higher benefit than is now possible for the first few weeks of unemployment. In accordance with the general principles which have been discussed, these lump sums or increased payments should not be attached to particular purposes, such as moving house. How they are used should be left to the person concerned. But as far as retraining is concerned this is a field where the State should collaborate with industry to provide the schools or courses. This brings me to the question of industry's responsibility for unemployment due to changed techniques. Compensation for loss of office is now frequently paid to directors. Why should not industry make similar provision for skilled workers? It should, I think. As co-ownership grows the need for such a provision will decrease. But in the immediate future at least, contracts of service allowing for notice of at least a month, and in some cases very much more should be encouraged and if necessary enforced by law. The State, which in this context means the general tax-payer, is doing a great deal for industry. I see no reason in principle why, in addition to the contributions of the employer to the weekly stamp, industrialists should not contribute to a fund for those with whose skill they find they can dispense. If automation grows this will become important. It may not be necessary to have any separate fund, and it may be more equitable to impose an extra payment on industry at large rather than on individual employers, for the small business might find it intolerable to have to bear the whole strain of a

THE SOCIAL SERVICES

sudden change in its methods, but the principle of some such payments seems sound.

Family allowances and food subsidies have helped to ease the strain on people who were in real poverty. The latter have almost disappeared and can now be dropped except for school meals. Some small element of subsidy is justifiable, partly for administrative reasons and partly to promote goodwill between parents and the education authority. There is also a case for assisting those schools where the expenses of providing meals is liable to be heavy. Family allowances are a more difficult case in the short run. I suspect that though one of the most illogical, they have been one of the most useful subsidies. After being opposed by organized labour they have come to be widely appreciated as a help to family income at a time when it is needed. On entirely empirical grounds they should be left.

Except in the very short-run the payment of housing subsidies regardless of need is indefensible. No Government can bind its successors, but at present the local authorities have some sort of promise that subsidies will continue for certain purposes for some time. In 1956 the Exchequer subsidy to local authorities was £63 million.

How far this promise could be limited in time or amount is a matter of consideration, but that in general Liberals are against the present type of housing subsidy should be made clear: whether this subsidy comes from central or local authorities.

As Liberal proposals for lower taxation, co-ownership and political reform take root the lesser social benefits should be allowed to wither away. There should soon be little need, if we advance at all towards a Liberal society, for death grants, etc. Higher incomes will make them unnecessary. The war and disability pensions must remain until we reach the stage of a national minimum income.

In the short run, therefore, we should aim at a simplification of the present services, reform of National Unemployment Policy and the gradual reduction of the services as personal income and saving are allowed to increase. Are there any new services which are needed? Bearing in mind the Liberal view of the test to be met before a new service is set up and the paramount need to check the extension of government action into inappropriate fields, I think there are few new personal services. The old are sometimes apt to be left out in the world of today. There may be some need for something more than their pension. But Liberals would be unwise to lay down any rigid procedure for dealing with the innumerable variations on the theme of old age. A higher cash pension would solve several of them. Old folks' homes are the only place for many old people, and a suitable place for some. But fundamentally this is a family problem. Better and more imaginative housing, higher living standards leading to more leisure, especially for women, voluntary services or even services by the State or local authorities such as the provision of meals against payment, above all, foresight by families themselves in preparing for the old age of parents and grandparents, could all help.

Perhaps, too, the time may come when we should do more to assure children of a start—a Liberal start—in life. Here, again, the rub is the family. I do not share De Madariaga's views on the family in all their intensity, but I agree with him when he says: 'Of all aspects of human environment the most important by far is the family. The family is the first cell of the Social Organism, the first institution which man encounters upon his arrival on earth, the institution without which he cannot physically survive. It is the family which makes childhood and adolescence possible to the being who is as yet incapable not only of rendering any service to society but even of ensuring his own existence. It is also the family which will teach him the traditions and culture which that part of the world where he is born has acquired through the centuries. True Liberalism, therefore, cannot attach too much importance to the family.'

Importance, yes—the family is very important. But De Madariaga and other Liberals sometimes gloss over its vices. The danger of attributing human characteristics to abstractions is perhaps even more misleading when done to the family than it is when done to the State, Government or the community—for it may appear to have more justification. The characteristics which must be attributed to it are by no means always benevolent. There is no greater bully than the family at its worst. Because the family is important that is no reason for not trying to improve bad families

THE SOCIAL SERVICES

or rescue some children from it entirely. The care of children from bad homes is a social service on which more emphasis should be placed. On this, I think, most Liberals would agree. Whether we should go further and consider the eventual endowment of children is a very much more open question. There is a considerable breadth of opinion among Liberals about death duties and inheritance. We should be united in saying that everyone has the right to leave a proportion of his property to whom he wishes subject to provision for his wife and children. But there might be a great many views about the size of the proportion. This is another matter which it will be much easier to handle when wider opportunities to accumulate property are offered to all. What we can say definitely is that we want to tax the size of the bequest—not only the size of the estate. But the problem will remain: we believe in some equality of opportunity: we make this the foundation of a first-rate education available to all: but the advantage of a child with a generous but thrifty father over one who suffers from a spendthrift drunkard will still be very great. I am afraid we have to fall back on the answer that in a human, changing, open society, we cannot eliminate all risks even for those to whom, as is the case with children, society owes special obligations. In a very rich society I, myself, would be sympathetic towards the lump sum endowment of totally impoverished children. This is, to my mind, the sort of social service which it might at any rate be wise sometimes to consider.

Do Social Services debase character? Do they encourage fecklessness, waste, the idler and the sponger? Possibly they do, but in a small degree. My experience is that there are deliberate shirkers who would work if they could not get National Assistance, and there are feckless people who nevertheless would save if they knew that otherwise starvation faced them in old age. But the net figure, so to speak, of such bad characters is small. Some would live by scrounging even if there were no social services. The main incentive to work and saving is not, and should not be, poverty. The old conception of the extreme temptation of idleness (peculiarly embedded in the poor, neither unearned income nor indolence were ever considered so wicked in the rich) has been largely disproved. But the type and size of the social services have subtler

THE SOCIAL SERVICES

effects. They create pressure groups and vested interests. They encourage curious and harmful attitudes. For instance, there is the claim of the old-age pensioner that he has a right to his pension but is insulted by National Assistance. This reflects a natural and honourable instinct, but is not strictly justifiable. In fact, very soon very few pensioners will have contributed anywhere near the sums which would actuarially justify their pension. The difficulty of defending the payment of 10s. to some widows is great. If they deserve anything, why not more? But in any event in what way is a widow with no dependants necessarily more deserving than a spinster of the same age? This sort of illogical special payment, difficult to defend but hard to withdraw, leads to feelings of unfairness, pressures to make further concessions which all too often make the confusion even worse and the division of people into 'pensioners', 'unemployed', etc. I cannot see that, with suitable safeguards, a national minimum income somewhat below the minimum wage, but above subsistence level, would be any more debilitating to character than the present services.

7

Education

It is not my business to say much about the mechanics of education. I am not drawing up a political programme for education. There is a great deal of agreement that classes should be smaller, school buildings better, teachers paid more. We all hope for these improvements. But there is not nearly so much discussion of the content of education. Indeed, except for spasmodic demands that more science should be taught, there is very little discussion in political circles about what education is for, or indeed what upbringing children should be given. I don't know whether this is properly a political subject or not, and I don't think that it much matters. What is certain is that it is a subject vital to the running of any country, as well as to the children concerned, and particularly important to Liberals. For if you are going to put great stress on the individual, if you are going to bring more responsibility on him, then his education is a matter of great concern.

Education should flow along natural courses. It should start in the family. As so many mothers go out to work this is difficult. While any Liberal must welcome wider opportunities for women, as more leisure becomes possible for mothers—and for fathers—I hope we shall restore family life. The best school is not as civilizing as the best family and the average school is a great deal less civilizing than a decent family. It is not only manners that the family can teach. Parents should interest children in a reasonable competence with their hands—in orderliness—in observation—in general knowledge and indeed in the three Rs. Further they should show them the importance of the family itself. Anyone who makes a success of family relationships must have had some training in ordering his or her life. But at present many parents are in-

EDUCATION

capable of giving their children a good upbringing. Some do not have room. Some are too constantly harried by their own poverty or are simply too ignorant or ill-natured to look after their children. It is a Liberal principle that grown-up people should be allowed to lead their own lives. It is not a Liberal principle that children and others who cannot fend for themselves must be left to the harsh mercies of their environment in the name of liberty. Even if a bad home is often better than a good institution, the State must concern itself with children who have no proper start in life. It can do so in a number of ways. It can ensure that children at some point in their education are taught their future duties as parents. Certainly the importance of family life should be stressed throughout the educational system. Instruction, too, should be given in all sorts of practical matters which arise in the running of a home and the rearing of children. Secondly, nursery schools and boarding schools should be provided for children who need them. Thirdly, the law should attach far more importance to good treatment of children. I know it is often said that you can't make good parents by sending them to prison. But often we read of appalling cruelty to children which draws a light sentence compared with what would be imposed for any comparable attack on a grown-up: it seems, too, that the law still regards offences against property as more serious than offences against children. I do not; though I concede at once that the punishment should be reformatory rather than retributive. Lastly, voluntary bodies should be encouraged to carry on home services. I hope that in a Liberal society, volunteers would be forthcoming. If they are not the State may have to organize a service. Services which help with old grandparents, or enable mothers to get some time off, may have a very good indirect effect on children by making their home less stormy or stressful.

When it comes to what is generally described as education, that is to say teaching in a school, I do not want to see this divorced from life at large. The great drawback of the British public school system has been that it creates a world of its own. The argument about how far education should be a lesson in how to earn a living and how far it is to be a more general introduction to art, literature and civilization seems to me a peculiarly British dilemma. In few

EDUCATION

other countries is art so sharply divided from ordinary living: in only the most advanced and puritanical countries are people bothered about Mammon v. The State of their Soul—which is often confused with their social status.

At all stages of education, emphasis should be laid on thick about doing the job well. Great satisfaction is to be got from doing the simplest actions efficiently. It is also very practically rewarding. I think we pay too little attention to mental and physical expertise. In the army I was much struck by the astounding variations among soldiers in their ability to fold up tarpaulins, or even their clothes, or find things, and by the variations of the time taken to wash and dress. We have tried to escape from some of the dreariness and monotony of old-fashioned education. But I agree with Bertrand Russell in thinking that a certain amount of monotony, of rhythm, of insistence on the same job accurately done should run through children's lives. We send small children to school for too long. The school periods can well be short and the subjects simple, but neatness, order and accuracy should be insisted upon. And while very small children should spend most of the day amusing themselves, they must be taught to read, write and do arithmetic. If this seems a glimpse of the obvious, we must remember that the number of boys and girls who quickly forget how to write neatly —in some cases to write at all—is alarming.

As soon as we move away from the hard foundations of any education the difficulty arises that children mature at different speeds. If arithmetic is carried on through primary education it seems to me that up to about ten or twelve the only other subjects which are good for all children, or nearly all, are history, geography, literature, art, one foreign language, and some practical instruction. A cursory examination of many school syllabuses (it is surprisingly difficult to find out what is taught in Britain today) gives the impression that a bewildering number of subjects are touched on. I believe that children are cleverer than they were (probably better feeding alone enables them to do more mentally as well as physically), but I find it difficult to believe that a high proportion can, especially in large classes, take in all that is offered them. Primary education should be simple. It need not take very long. If the school-leaving age could be delayed I should be happy to see

EDUCATION

a later school-starting age. But there should be some test, some examination to find out what the child knows. The present eleven-plus examination is largely designed to find out, not so much what it has learned, but of what it is capable. It is an intelligence test. Like most intelligence tests it finds out one sort of intelligence: that may be as much as we can hope for in such a test, and a test of the child's potential may be necessary; but it is not the same thing as an examination in certain subjects.

Having passed an elementary examination what should the child do next? Here we come against the great difficulty of general education. A child of limited intelligence and small intellectual enthusiasm may not get a great deal out of education between the years from say twelve or thirteen to sixteen or seventeen. Yet there are things which it should be taught, but which are difficult to teach to such children until they are older than the present school-leaving age. The important thing that you can teach a child—and the one which is much neglected—is to think logically, to order and marshal its capabilities, and to use its judgment correctly. This is important in almost any job, certainly important in life generally, and vital if you expect to run a Liberal society. When is this to be taught? When are children to be practised in using their minds, in discriminating between true and false reasoning and in following an argument? It is said that this process, which is sometimes called mental discipline, is the purpose of studying Greek, Latin or mathematics. This may be so, though what I have in mind is not precisely the same as mental discipline. There are several eminent scientists and politicians for instance, whose mental powers I would not presume to criticize and who have survived stiff tests in mental discipline, who nevertheless are capable of talking good old straightforward nonsense about many subjects. More serious, however, is the objection that there are a great number of children who are incapable of getting far with Greek or Latin or mathematics before they are eighteen (even if then) and who will have left school before they derive any benefit from these subjects. The solution is to make the training of the mind continuous. From the moment a child learns to read it should be taught to concentrate and to reason. There must be no sharp break between primary and secondary schools. At present it does

EDUCATION

not seem that much stress is put on training in orderliness and judgment. Vocational training; social training; education for leisure; development of taste; awareness and understanding of the world; all these are mentioned in syllabuses I have examined, but there does not seem much effort to stop the kind of argument which, for example, says, 'Prices have gone up, therefore I must have more wages, and if that puts up prices further, I must have more wages still.' In the syllabuses of one school I read this: 'It is realized, of course, that *some* logical order must be observed in teaching mathematics, even if that order is only present in the mind of the teacher and is not apparent as an appeal to pupils.' The logical order in mathematics must surely be apparent to the pupil as well as the teacher unless the subject is to have no value at all as a training. If a child is incapable of appreciating the order in mathematics before the time comes for it to leave school it must be given a chance to revert to education when it is more fit to be taught. Liberals would insist that further education should be every bit as serious as basic education. Now it seems too often to consist of a very small pellet of knowledge in an inordinate amount of jam, or of subjects like gardening, bird-watching or hat-making. These have their place, but there should also be a place for the man or woman who at eighteen or twenty, or later, finds that he or she can understand things which were gibberish at an earlier age. However, I return to this later on.

In the meantime we must first consider secondary education. Among the rich the break comes at about thirteen when the child moves from preparatory to public schools; among those educated by the State it comes rather earlier. At present children between the ages of twelve and fourteen are, very roughly speaking, educated as to 60 per cent at secondary modern schools, 17 per cent at local authority grammar schools, 2 per cent at local authority technical schools, 1 per cent at comprehensive schools, 10 per cent at various other local authority schools, and 10 per cent at independent and direct grant schools. The great majority of children leave school at fourteen or fifteen. I am not convinced that there should be this separation of the sheep and the goats at eleven-plus. Liberals must give thought both to the timing and the nature of this break. Most people have some sympathy with the idea of a comprehensive school, particularly in rural areas, if the disadvantages—which seem largely administrative and not educational—can be eliminated. But if it is a good thing to have a school which comprehends the clever and the stupid (as Eton used to do and to some extent still does) why not run the present primary and secondary schools together? If we want to maintain continuity in education there seems a lot to be said for running the syllabuses through from school-joining to school-leaving age. No Liberal wants to destroy the grammar schools, but if only some 15 per cent to 20 per cent of all children go to them there is plenty of scope for improvement in the education of the secondary modern pupils comprising 60 per cent of the rest. As far as these children are concerned, I see no reason for making the break, and the only break, at eleven-plus. As time goes on we should encourage the 'growing-together' of education. All children need some vocational training and some general training, some technical education and some 'education for life' at least up to fifteen or sixteen. At present there is, of course, a very steep fall in the numbers of children at school at all after fourteen and in every subsequent year it falls again. Of every 1,000 children who receive any sort of secondary education only some 350 are still at school when they are fifteen and a half.

But this programme running on to fifteen would be open to two objections. To have boys and girls from five to fifteen in one school might be difficult: and with the present pattern of education it would be troublesome to run primary and secondary schools in this way. But let me repeat the suggested scheme. Liberals should aim at the first stage of education at home or in a 'nursery school'. This might take the child up to five or six. The next phase would last to twelve or thirteen, and should allow some choice of school. There would be no one rigid test, but a series of examinations to determine how much had been learnt, and how much was still to learn. In the third stage, the greatest possible variety of educational food should be offered. And this is the stage which might present difficulties. But in the long run it does not seem to me to matter so very much in what type of school this food is dished up. In fact, the more types of school the better. What we want to aim at is education up to sixteen or seventeen years of age in schools which may pursue different methods and curricula, some putting rather more emphasis

EDUCATION

on one subject, others offering their own traditions, but all giving some education in the arts, and in science. Education must not be forced into a set pattern. What is certain is that to provide any sort of a system in which serious 'secondary' education is attempted for all, the normal school-leaving age must be raised, though for my part I would always be in favour of leaving it flexible within limits. It seems to me that if child, parent and schoolmaster are all agreed that the limit has been reached at, say fifteen, then the child should be released—especially if by then we have a better system of further education through which the child may get back into the stream of teaching if it develops a thirst for knowledge later on in life.

I have said that it does not so much matter in what sort of school a child is taught as what it is taught. Each sort of school has its advantages and its problems. It is said that to get a sufficient proportion of clever children in a comprehensive school to justify a teaching staff for the higher levels of education you have to have an inordinate number of pupils. But in time we must hope to raise the general standard and to produce more sixth-formers. In time, too, all education after the primary stage must include arts as well as vocational training. This is already happening. As a report of one secondary modern school says, 'There is, however, an insistent and almost unanimous expression of opinion amongst employers that what they look for above all in modern school-leavers is evidence of sound general education.' Thus we should aim at making the 'comprehensive school' smaller and more like the grammar school: or we should group various school buildings together or institute the house system as at Coventry.

As to the independent schools, including the public schools, any Liberal view of education must allow them a place. Liberals, above all people, and in education above all subjects, must not allow convenience or symmetry to dictate uniformity. It is quite untrue that the public schools must either be abolished or wreck the whole system. Let them continue, indeed, let tax relief or grants be given to parents who want to send their children to them. If public school headmasters are sensible, which they are, they will try, once the handicap of having to find the full fees is removed, to broaden their recruiting base. They will try to go be-

EDUCATION

yond the Fleming Scheme and take a far bigger proportion of poorer boys. If they fail to offer an education which, to a considerable number of people at least, is better than the publicly provided school, the public schools will wither away. I am not at all sure that this will not happen. Whether it would be a good thing if they disappeared depends very largely on the reason. If they are taxed out of existence while the education they offer is manifestly better than public education, it will be a bad and illiberal thing. If, however, public education becomes so good that no one wants to send his children to Eton or Winchester, it may be a good thing. For, in my view, day schools have many advantages over boarding schools and the virtues, like the vices, of public schools have been exaggerated. The main fault of the public schools at present is that the high-taxation-socialist type of State has had the paradoxical effect of making them the perquisite of the rich alone. Before the war there were a number of comparatively poor families who could afford to send a son to a public school. These have been largely eliminated, not by the public schools, but by Socialism.

There are always a few children, whose parents are dead or abroad, or who come from bad homes, or who are better away from home life. There is sometimes one cleverer or stupider child in a family who will be better educated at a boarding school. In the course of time it might be a good thing if boarding schools were provided for them by public education authorities, if the public schools do not meet the need. Already in some places where the distances are great, hostels are provided for secondary pupils. While on the whole preferring that the greatest contact should be left with the family, on the grounds that the family is normally more civilized than a school, nevertheless where this patently is not the case, Liberals must allow that boarding-school education has advantages.

What is to be taught to boys and girls in the thirteen to eighteen group? I again reiterate first logic or straight thinking. But after thirteen it should be possible to relate this to rather wider subjects. Already at most secondary schools there is instruction in how government works and in the duties of a citizen. There is always a fear among Conservatives that if this is pushed too far it becomes tendentious and harmful. Liberals should not be put off by

EDUCATION

these fears. Of course, all or any education except the most technical may be tendentious; most harmful of all perhaps is the sort of teaching in history or politics where the bias is hidden. Democracy depends for its survival on a proper appreciation of how it must be worked. We cannot take it for granted that every boy and every girl is born a little democrat. Nor can we assume that the principles of democracy will somehow percolate into general political practice without constant restatement. Generations are growing up who know nothing of Mill, not to mention later political democrats. It is not nearly enough to tell children how Parliament or local government functions. They must be taught to face the difficulties of political organization, encouraged to be critical, especially critical of our success in giving effect to the principles of democracy. They must also be shown the principles and effects of other political systems. The safeguard against undue bias is that varying views should be explained to the child—not that a veil be drawn over dangerous subjects. The presupposition of democracy, the differences between Liberalism, Socialism and Communism, the need for critical examination of all the catch-phrases of politics, all this must be opened for children at some stage of their education. Though I realize that many children will not make much of it before they are sixteen or seventeen, nevertheless something should be attempted even before the school-leaving age is reached.

Assuming that reading, writing, arithmetic and an outline of history and geography have been grasped, and that the children have been introduced to literature and art, then science, or rather its principles, seem the next most important subject for the 'secondary' phase in education. Many people have been talking about the need for more scientists and much alarm has been sounded at our small output of scientists compared with Russia or the U.S.A. Liberals share some of this alarm. Though education must be a training for life, it must also be a training for a job, and though no one wants to force boys and girls into jobs for which they have no inclination, nevertheless it is fair to warn them where the best chances lie and to show them the advantages of certain careers.

It seems certain that the demand for scientists will be strong and prolonged. We have seen the predictions of almost unlimited demands for food and coal partially falsified in an astonishingly short period. But bearing in mind that all countries are trying to make themselves more and more scientific, it looks as though the world demand will continue for some time. But when it comes to deciding what sort of scientists are wanted, then the prophets should be wary. The Minister of Education spoke this year (1958) of increasing the annual output of scientists and technologists from its present level of about 11,000 to something like 20,000 by 1968. The Advisory Council on Scientific Policy has spoken of this latter figure as a minimum. But the word 'scientist' or 'technologist' covers a multitude of skills. It seems to me impossible to forecast what sort of scientists, engineers, skilled men of different kinds may be needed as techniques change. It may well be that some changes of technique will very much reduce the amount of skill required in various industries. Automation itself is doing this in some processes.

We shall need more teachers of science and technology. There will always be a demand for the first-rate or original scientist. It will be an advantage for any young man or woman to understand scientific methods. But further than this, scientific or technical training will have to be very adaptable. At present only a small proportion of children go on to technical colleges. In an ideal Liberal world these colleges, greatly expanded, would be on 'the campus' of universities but free to indulge in very considerable specialization providing that their students had a thorough general education behind them. They would, in fact, be rather colleges of a university than secondary schools. In the secondary plan itself from twelve or thirteen to sixteen or seventeen, all pupils should be taught some basic technology and scientific methods. Exactly how far children can be taken in science before seventeen is the sort of decision which must depend on the child. If we maintain a Liberal variety, not only of schools, but of courses within most schools, the emphasis can be changed from arts to science according to aptitude.

For those children whose bent is not scientific, the choice of subjects is more difficult. The science specialist is seldom totally allergic to history, art, music, religion, languages or literature. Even if the emphasis in his or her course is on science and tech-

EDUCATION

nology it should not be difficult to devote an adquate period to what is often called 'education for leisure'. And the child's brain will be adequately stretched by the main scientific subjects even if it does not get very far in the arts. The danger on the arts side seems to me that while the intelligent child can get a great deal out of, say, history and literature, the less intelligent child is apt to have unduly low standards accepted.

There is the dilemma for the less intelligent child between allowing it get along fairly well in subjects which do not overtax it, or more ambitiously trying to make it do work in which it may completely fail. In some of the pre-war public schools where a good many pupils were not pushed at all after the age of about sixteen, they became history or English specialists, they acquired a certain facility in writing essays and were left to an increasing extent to their own devices. If Latin and Greek are completely dropped from most curricula, as they are now, and in my view with good reason, arts education may be deprived of its main stiffener.

One way of mitigating the danger is to insist that some serious scientific training is given to all arts specialists. Another is to bring in economics and philosophy—at least political philosophy. How this can be done without overloading the curriculum, or boring the child into mutiny I do not certainly know. At sixteen or seventeen even clever children have a limited capacity for understanding Dicey, let alone Kant. At present while teachers are very conscious of the need to cover a wide field the 'A' streams seem to be very specialized, while the degree of specialization decreases in the lower streams. A teacher naturally wants to turn out successful pupils and as a result may not press a child to specialize when the results are bound to be mediocre. One of the as yet unproven fields is how far with improved teaching methods and ever increasing educational opportunities, it is going to be possible to get more children to tackle more advanced stages in all subjects before seventeen. To judge from the syllabuses of both grammar and secondary modern schools, a great range of subjects is already taught. And to the parent, at any rate, the questions in the Higher Certificate Examination look difficult. No great fear need be felt that the clever child is under-exercised mentally. It is the less clever which cause anxiety. We must ensure that these, too, do not leave school

EDUCATION

without ever having stretched their brains, unpleasant as the process may be.

I come now to the universities. At present only about 5 per cent of children go on to a university, and some 12 per cent to training colleges of various types. This is too low. The difficulty of getting into a university also seems to put a premium on examination-passing which is a great strain on many children. When I emphasize the need to stretch a child's brain, I do not mean merely that a high premium should be put on exam-passing. Examinations certainly have their place; like the prospect of being hanged they wonderfully concentrate the mind. But on some children—girls especially I suspect—they put an undue nervous strain. Liberals would like to see the number of university students increased. How to do it? There is no one single solution. It must be done in various ways. Oxford and Cambridge might be expanded still further, but the limits of expansion seem near. On the other hand the Scottish universities could be very much enlarged. One of our mistakes has been the comparative neglect of St. Andrews and Aberdeen. Here are universities in being with the right traditions and surroundings which could be greatly increased in size. They could also absorb some of the overseas students whom we are now forced to turn away. One of our major insanities is our failure to cater for students from the Commonwealth, the Middle East, and all Asia and Africa. It shows a lack of imagination on our part that Oxford and Cambridge, and to a lesser extent London, have been allowed to usurp the image of a university. To create a new setting for a university in an industrial town may be difficult, though by no means impossible, but in St. Andrews, Aberdeen, Edinburgh and even in Glasgow, we have a nucleus which cries out for expansion. Alas, expansion is nearly always in the form of ugliness built, with that unerring eye for destruction which characterizes the British, upon the ruins of beauty. From all the wastes of dull and indeed slum-ridden Edinburgh, for the university to select George Square for destruction, passes the understanding of anyone who has not studied the activities of academic institutions.

The first thing, therefore, that Liberals have to say about the extension of the universities is that we should see what can be done

EDUCATION

with those we already have. Nor need this be confined to the older foundations. 'Red-brick', too, can be enlarged and improved. Not only enlarged and improved, the 'provincial' universities can also be diversified. The levelling of instruction and grants is all very well up to a point but beyond a certain minimum equality here becomes as deadly as anywhere else for it comes to mean sameness. Ideally a university should be a universal teacher: it should teach all subjects and cover all knowledge. But in practice this can hardly be attained. Nor is there any reason for complete universality. There is no need for Exeter to teach medicine as long as there are adequate facilities available to local children elsewhere, nor for Hull to teach nuclear physics. The purpose of teaching or studying the higher reaches of knowledge at a university instead of a technical college are threefold. First to assure that some background in civilization generally is absorbed at the same time, secondly to provide that teachers and students meet their opposite numbers in other subjects, and thirdly to provide an agreeable life: a university is not merely a means to an end, it is an excellent thing in itself. These aims can be reached without turning every university on the same pattern. Some can ignore a subject entirely, others can put greater emphasis on one group of studies, some may offer degrees after a fairly narrow course of instruction, others may develop courses like P.P.E. at Oxford which includes several subjects. As a corollary to this they should be allowed to vary the salaries offered to professors and lecturers.

But even if we expand existing universities new and old, this will not be enough. I do not believe that any Liberal can be dogmatic about who should or should not go to a university. The cleverer children should have priority and there will always be some who neither want a university education nor would benefit from it. But the rich have never believed that education should go to the clever: and on the whole they were probably right. There is a lot to be said from everyone's point of view in having some average students at a university. A distinction really falls between those who can benefit from some part of university life and those who cannot appreciate it at all: and this distinction is not the same as that between the clever and stupid. The idea of second- or third-class universities—county colleges—and so on, repels some people because they see in it another step towards segregation. The less clever, already condemned to 'lower' secondary schools and 'inferior' streams will be faced with an education of prolonged dreariness. This fear could well be realized. But it need not. We ought to found many new types of advanced colleges, grouped together so that they can find a focus like a university. We should not be afraid of starting less ambitious institutions for educating students over seventeen so long as we allow them to grow, so long as variety is allowed. In America there are innumerable universities, founded by private individuals, states and churches. Some, no doubt, are not very good. But nearly all do more good than harm, and the gradation from some small middle western college, little more than a rather agreeable school, to Harvard or Yale is so gradual that there is no division of 'sheep' from 'goats'. I would like to see the same here, colleges of every sort, technical, agricultural, colleges devoted to the arts, non-residential colleges, grouped together where possible and offering to at least 25 per cent of the population the rudiments of a university life.

Finally, there is the education of people who have left school or university: further education as it is called, though that term covers all sorts of activities which are recreational rather than educational. In rural areas, in particular, further education could be a great boon. The present educational system does not work well in thinly populated country districts. It is bad enough to close the 'side' or village schools. But it is the secondary phase which takes children out of their environment. The bad custom which has grown up of regarding the secondary modern schools as 'inferior' is at its worst in the country. For the mark of failure put upon children who don't get to the grammar school, is associated with staying at the only type of school the villages offer. It becomes a matter of prestige to leave home. It is obviously impossible to offer a grammar school standard of education in every parish. Further education cannot completely overcome the difficulty. But it can help. Suppose we reached a stage when education for all was completely compulsory up to twelve or thirteen, then a new phase began lasting three to five years according to the child's capabilities, its wishes and those of its parents, and the sort of employment which it was likely to undertake. In some country districts this further

EDUCATION

phase of education would be carried on, for some children, by travelling teachers taking small classes in the evening, or at least for half the day only. It might well, too, be of a largely technical sort. There will always be some children who at about thirteen or fourteen get very restive with school but who are capable of making good farmers, mechanics, craftsmen, even though they may not be so good as to justify sending to a technical college. For these a mixture of work and education, rather on the apprenticeship model, seems appropriate. Courses run by the education authorities in conjunction with the agricultural colleges and local farmers would meet the need for effective education in scattered districts. There is always the chance, too, that someone who was profoundly uninterested in learning at, say, fifteen, may develop an interest at twenty. If some opportunities for further education exist these requirements can be met.

But of adult education I am sceptical. The intense desire for knowledge which existed fifty years ago seems to have evaporated or is now satisfied by TV and the papers. By all means let adults come together voluntarily for education; by all means let the universities run extra-mural courses, but modern adult education of people over twenty-five is largely a matter of providing 'third programmes' in various ways outside the educational system itself.

Fifty years ago any Liberal politician writing about education would have discussed religious teaching at some length. Today it does not seem to present such great difficulties. But there are still differing points of view on the teaching of religion and morals, and to a Liberal these are topics of importance. I am not greatly perturbed by the prospect of non-Roman Catholics, non-Episcopalian or Presbyterian taxpayers for that matter, being forced to subscribe through taxes for church schools. The Scottish attitude to Catholic schools seems a sensible one from the practical or financial point of view. I am much more perturbed by the demand of the Catholic Church for absolute monopoly over the moral and spiritual teaching of Catholics. As I have said, I am strongly in favour of children being introduced to 'dangerous subjects' in politics or morals, but they must be allowed to look at both sides of questions while being constantly instructed how to make up

EDUCATION

their minds eventually. If the Catholics or any other body could impose an effective censorship on all other philosophies than their own I should be against allowing them to retain the children of Catholic parents entirely in their own hands. But as in practice they cannot do this, and few of them are so bigoted as to want to do it, the Liberal attitude should be to allow denominations, or indeed other bodies, to found and run schools with Government assistance so long as they come up to a certain standard.

In State schools I would not only allow but encourage the teaching of religion and morals. These should be matters of peculiar interest to the family and therefore parents must be allowed a very considerable say in what is taught. In a Liberal society they will have some choice of school and therefore some choice of avoiding those teachers with whom they may violently disagree. But this choice will always be variable in amount and in some places may not exist. A strong obligation therefore rests on school-teachers to consult with parents. The system by which ministers of different denominations come into schools and teach such pupils as are of their church seems a good one. But we have erred in trying to limit the moral and religious teaching given by teachers themselves. So long as it is open to correction and given sincerely and with the approval of parents, the only safeguards I would establish are that such teaching should be suitable for the age of the children concerned. If Liberals are then asked if they would allow Communism or free love (whatever may be understood by that) to be taught, I think we must squeeze the emotional bias out of the question. Liberals are rational. They believe in Liberalism, which is essentially bound up with democracy and personal morality, because they think it right, morally right and practically correct. They believe this can be demonstrated to the reason. Therefore, if Liberalism is explained as well as Communism they have nothing to fear from the latter. In fact, Marxism must in Liberal eyes be explained at some stage. For one thing it is an important strand in European thought and without it modern politics cannot be understood. Secondly, Liberals can learn from it. Thirdly, it is the mainspring of the Russian revolution and without some knowledge of it we shall not understand half the world. Therefore, while it should not be taught to children of six

EDUCATION

and no monopoly should be allowed to Marxist teaching, to forbid it to be mentioned would be absurd.

Matters of personal morality may seem more difficult. But no Liberal is going to suggest that schools should be a-moral. If morality is to be taught as it must be, it cannot be taught in a Liberal society simply by stating it as an unchallengeable dogma. Any ethic must be open to discussion and the opposite point of view stated, if it is to be rationally accepted. Again, I am not suggesting that a full discussion of, say, sexual ethics should be carried on before classes of young children. At appropriate times there should be no ban on such discussion by qualified teachers with parents' co-operation. Many of our political troubles, e.g. race feeling, might well be cured by a more definite teaching on moral subjects.

The staffing of schools, colleges and universities which would come near providing the ideal education for a Liberal society would be expensive in teachers, in money and in real resources. I will not discuss money or school buildings for the reasons I have already given. But expenditure on education is a type of expenditure of which Liberals must approve. As always, they see no reason why some contribution might not eventually be met by the families. But children being unable to fend for themselves are a proper subject for the help of society, given through the State. The supply and training of teachers certainly needs more examination.

In 1958 there are some 260,000 teachers in maintained and assisted schools. Some 32 per cent of junior pupils were being taught in classes of over forty and some 62 per cent of senior pupils in classes of over thirty. There is already, therefore, a shortage of teachers. If classes were to be reduced in size and the school-leaving age put at sixteen or seventeen—not to mention the demands for new universities and technical colleges—there would need to be a very large increase in the number of teachers. At a guess comprehensive education might need 400,000 teachers if it were to begin at five. Could even the most ideal Liberal society find or maintain such a staff? I think it could, but it might well be possible to make a desirable saving in valuable teacher-power by raising the start of primary education and making more use of

EDUCATION

nursery schools. For some reason the notion that women with a different sort of training and lower academic qualifications might give very good service, probably in part-timé in nursery schools for children up to seven, is apt to arouse fury in the breasts of teachers. I do not know why. In the early stages of education the periods for which children can be taught are very restricted. The rest of school-time is already taken up with games or such-like activities. No one except a bigoted teacher can really pretend that this is the same sort of job as teaching physics to pupils of sixteen or philosophy at a university. To say that is not to say that teaching in a nursery school is an unimportant job, only to say that it is different. Perhaps it would be as well not to call these five to seven or eight schools nursery schools. But whatever they are called they are not the same as schools for older children.

If we are to get the teachers we must pay and train them. At present too few teachers have a university education. If any specialist colleges need to be associated with universities they are the teachers' training colleges, the ultimate aim being that every teacher should have a degree. But even if this were possible a variegated programme of education is going to make great demands on teachers. Already there is the danger that we shall draw up syllabuses and programmes which exist on paper only. Education in the end depends very much on the quality of the teaching.

I have tried to deal in this chapter first with the timing of education, making some suggestion where breaks should come and where and how it should progress and end. Otherwise, I have been largely concerned with what is taught. I believe that these are the aspects of education which will remain fundamental. Though we can be dogmatic about very little else in education we must be dogmatic about its importance and about the supreme importance of teaching people to think and to think not only about the safe but also about the dangerous subjects of life.

8

The Liberal Environment

A civilization is judged by many tests, but not least by its contribution to beauty and knowledge. The Greeks are forgiven their slavery, the Renaissance despots their cruelty, because of the beauty they left behind. It is sad to think that when hundreds of years hence people read of the great empire of the British they will be able to find very little in the visual arts by which it can be pleasurably remembered. In our colonies and dominions we have left some heavy government houses usually looking like Victorian villas blown up ten times life-size, and a few barracks usually designed for somewhere different from their eventual resting-place. There are, I believe, interesting relics of eighteenth-century architecture dotted about our dominions, but little else of note. As for statuary, streets or pictures, fine work may be seen in Melbourne or Ottawa, but it will almost certainly not be British from the great British pro-consular age. New Delhi is perhaps an exception but, if it is, it is something of a farewell gesture. I am, of course, talking of what the native British have done: the Australians are producing artists on a scale out of all proportion to their population—why, no one seems to know.

This should concern politicians and, above all, Liberal politicians. It is disquieting that we should compare so unfavourably with other ages and indeed with other countries. The slums and vast ugly cities bequeathed us from the nineteenth-century are blamed on private enterprise. But, if anything, our recent architecture is worse than Victorian—at any rate it is a serious rival. Our lamp-posts, our cinemas, and, it must be admitted, our modern university buildings reach a standard low enough for any jerry-builder of the last century. There are few parts of the world so ugly

THE LIBERAL ENVIRONMENT

as the Midlands of England, unless it be the Midlands of Scotland with those new housing estates of mud-coloured four-house blocks, ill proportioned, dull and ubiquitous. Acts are passed, books such as *Outrage* are written, the *Observer* denounces, but the blight goes on.

Drama and literature are in much better shape. But they, too, raise questions for a Liberal politician. They seem unable to support themselves but are they suitable subjects for state-aid? How does art meet education?

Round the periphery, so to speak, of art proper are all the advantages of civilization: newspapers and wireless, social life, holidays, leisure, sport. What would Liberal civilization be like, how would it be managed?

I am even less qualified to answer these questions than many others in this book. But there has always been a connection between the word Liberal and the arts. Any political creed, too, should be sensitive to its effect on civilization. Politicians should not attempt to make people aesthetes any more than they should try to make them good. They should, however, create the conditions in which the arts and civilization generally can prosper. Further, they should try to set an example.

Public authorities of one sort or another, local, national or international are going to build many buildings, roads, lamp-standards, bridges, telephone-boxes, etc., and they are going to furnish offices. Just as I hope that education and freedom of choice will cajole more individuals to prefer quality to equality so I look for some expression of standards, some personal taste from public bodies. This does not mean vast extra expense. On the contrary, we should save a great deal of money by ceasing to prolong into the depths of the country the trappings of the town in the form of concrete kerbs and asphalt pavements along every main road. We could save on less pretentious public buildings. We could save by insisting on variety. It is the sameness of public works (so appropriately and anonymously called) which spells death: the dreadful deadness of telephone-boxes, the same bogus Georgianism of which rises on the moors of Shetland as on the crossroads of London; the uniform drabness of government offices in which furniture, pens, ink and linoleum appear to have been

THE LIBERAL ENVIRONMENT

ordered by mile or gallon by people oblivious of their surroundings. Every now and then a minister or a public body attempts to break away. But nearly always they turn to the most academic designer or they form a committee. Taste, alas, is not an academic matter (look at the taste of university dons): nor is it a committee matter. The first step in public civilization is a revolt against conformity. Let Scotland and Wales, Lancashire and Sussex for that matter, design their own 'works' and paint their own colours. The next step is a revolt against anonymity. The names of ministers and their advisers ought to be fixed to the public buildings for which they are responsible. The drabness of the new Sheriff Court in Edinburgh should be a reproach for all time to its sponsor and architect: their names should be engraved on it. What we require of public authorities is not patronage of the safe, but some attempt to lead, even if sometimes they lead us up blind alleys. For instance, the Ministry of Works looks after a number of ruins: must it stick the same notice on every gate from Land's End to John o' Groats regardless of place or landscape? Must it cover its paths with the same chips in Durham or Caithness? It would be inconceivable if it did not happen. And the sacred cry of economy is often claptrap: the Highlands are dotted with imported prefabricated community halls and houses, much more expensive and much less satisfactory than those which could be built of local material by local builders. Public authorities should be trying to stop the appalling onset of what has been called subtopia: they should be giving rein to local or individual taste even if it turns out badly but, alas, they are the subtopians-in-chief. It isn't impossible, many of the greatest buildings of the world, the Parthenon and Chartres Cathedral among them, have been built by public bodies. Away with defeatism. Revive the standard of the Medici.

When it comes to creating the climate of civilization, the task is at once more difficult and easier. It is more difficult because anything would be better than the present furnishings of government offices yet there they are with our masters sitting in them: it is surely inconceivable that there are not in the Government service men who wince when they see such triumphs of public taste as the pit-head baths of peeling plaster and cracked concrete, pathetic

THE LIBERAL ENVIRONMENT

attempts at Mediterranean modernity in the climate of Scotland or Lancashire. Public buildings and furnishings are so low that almost anything would be an improvement. But when it comes to buildings, books, pictures, furniture for everyone, then we are at the mercy of what the fates provide us in the way of artists. Though architects complain bitterly of their patrons, they, to take them merely as an example, inspire little confidence that with the best patrons in the world, they would produce a great flowering of beauty. And the British are not the best patrons. They have long taken a delight in destroying beautiful things. But the task of creating the climate is easier perhaps because the comparatively few people, who in any society set its artistic standards and enjoy its beauty if given a chance, will probably tackle it more avidly than public servants recruited with no thought for their taste. If public servants cannot create the right climate they may at least allow a free hand to the few daemonic characters who mind where and how they live.

But how is the chance to create beauty or to appreciate it, to be given? We have heard a lot about the rich patron, and the need for the Government to take his place. I am not pessimistic, however, about the chance of a fair number of people wanting to make and possess agreeable things, if under a more Liberal dispensation they were left with more money and greater opportunities. We must revolt against the immense prices put upon fashionable works by a few well-endowed galleries and a handful of millionaire investors. We must stop prestige buying, or the filling of galleries with works for the benefit of curators and art students. Of course, the national collections should from time to time be enabled to buy more old masters or impressionists. But now that travel is much more common it is unnecessary that every gallery should consider it essential to show the whole gamut of art. They should be first and foremost patrons of living and national art (here is an appropriate and Liberal field for nationalism) and by so doing show to their visitors that pictures of their contemporaries can be rewarding. The hundred-guinea picture should be as common in people's houses as the hundred-guinea television set. But the concentration on collecting £25,000 pictures for public exhibition is killing the market.

THE LIBERAL ENVIRONMENT

Literature has not suffered the demoralization of architecture, nor, thank goodness, is it a good investment. But imaginative literature, the novel and poetry, seem to be rather in eclipse. This may be temporary: it may on the other hand be a natural development when television takes the place of a Scott or Dickens, Hardy or Conrad. In a perfect society, nevertheless, I cannot but feel that imaginative literature would still have its place, especially in Britain whose artistic fame rests on her literature. We could at least reward writers more generously. Although the importance of library buying is exaggerated, libraries supported by public funds are expanding—and usefully so. Even if more people have more money to spend the library has come to stay. It should contribute more to the writers on whom it depends. I see no reason why a small, a very small, tax should not be levied on every occasion on which a book is borrowed from a public library, and paid over to the author. I believe this is now done in Sweden.

Not only in modern times nor solely among the British has it proved impossible to pay for orchestras, opera, or indeed the theatre, solely on what can be charged at the door. Continental opera has always been supported by grants from governments or municipalities. Today the British pay relatively little out of taxes for the support of the arts. On what for some people would be pure Liberal doctrines we should not be taxed at all for these purposes. No one can say that opera-goers are the poorest section of the community and there seems no reason in theory why those who want highbrow entertainment should not pay for it. Those who like jazz or more modern forms of music find the money for them, which may make the music more original and the entertainment more enjoyable. There is no lack of interest in promenade concerts or Shakespeare plays. Should not a Liberal society be prepared to found and finance its own theatre or music? I doubt if a conclusive answer can be given to this until a Liberal society is in being. Such a society would have its own education, with emphasis on the arts; it would have, as I shall try to show, a more positive view of town planning which would provide at least the buildings and the environment for the arts; it would have more money in the pockets of more people out of which they could voluntarily finance the arts: all this would make State aid less necessary.

But, nevertheless, I believe a Liberal society through some organs—not necessarily the Government alone—should find money for the support of art. I reject some of the reasons advanced against this. I do not believe the public bodies are inevitably worse patrons than millionaires. Their record in architecture is bad partly because they have set themselves the wrong standards and have been dogged by a passion for uniformity, partly because good architects have not been available in sufficient numbers. They are not particularly successful as entrepreneurs of new artists or artistic developments. But the Arts Council through many activities and the City of Edinburgh through its Festival have at least been as good as many nineteenth-century patrons of the arts. Secondly, I do not accept the view that it is impossible for a public body to decide what is good or what is bad in music or theatre. I think the Third Programme did it well. But I do not indeed think a public body need make this decision. It can in the fields of music and drama offer facilities for a wide range of programmes on which there will be broad agreement that they are worth giving even if opinions will differ as to their relative merit. It certainly is not the business of, say, a municipality to educate its ratepayers in Beethoven. But there is a strong case for its subscribing, as many do, to the maintenance of an orchestra which makes it possible to hear Beethoven played. And the justification for this is that an orchestra is not something which individuals can support. Nor does it seem to me any more reasonable to expect those who go to its performances to pay the whole of its cost than to expect them to pay for the dustmen or soldiers: these, too, could in theory be supported voluntarily by individual effort. To enable us to accumulate a surplus for public support of the arts ought to be part of the aim of an expanding economic policy.

Liberals should welcome the development of the Arts Council and other such bodies and should be generous in their support. But they should give considerable thought to the methods by which they should be run and ideally they should encourage such bodies by educating public taste to dispense gradually with official financial support. One way in which this can be done is for these bodies themselves to co-operate with schools and universities in creating a taste for the arts. There should, too, be more participa-

THE LIBERAL ENVIRONMENT

tion by the public in the running of such bodies. At present the actual appointments to the Arts Council or to the governing bodies of, say, Covent Garden and Sadlers Wells are on the whole admirable, but the methods of appointment are unknown to the public at large. On the whole, appointments are made by a fairly small clique from a limited field. This is the case also of appointments both to the governorships of the B.B.C. and to the directorships of wireless and television programmes. I doubt if this is a field for democracy pure and simple, but it does seem a field in which some body appointed by other public bodies and open to representations from the general public might play a useful part. Lastly, though such undertakings as opera and ballet must to a certain extent be concentrated in the capitals and larger provincial towns, the provinces and small towns are too much neglected at present. It is true that the provinces are not always keen at attending or paying for highbrow entertainment. But the field should still be ploughed. From time to time companies are sent on tour, wireless and television has made it possible for people even in remote areas to hear good music, yet on the whole the concentration on London has led to apathy in the country at large towards the work of the public bodies supporting the arts.

I am well aware that these suggestions are tentative and open to objection. But this is a field for flexibility and for special engineering to meet special needs. If a John Christie appears he should be encouraged and if necessary financially supported; if the support is forthcoming from a John Lewis partnership let us welcome that, too. If no John Christies appear then some body, perhaps a body judiciously rooted in Nationalism like the Saltire Society, perhaps a voluntary body with specialized interests like the National Trusts, perhaps a larger conception like the Arts Council, must step in. There is an effort in this field to escape from bureaucratic control and it should be encouraged in every way. Here, too, there can be effective collaboration between public bodies and voluntary effort. In the end it is the artist who counts. A good society, a good Liberal society, will treat its artists well and give them honour and prestige. No one wants them to become self-conscious. But it is they who should determine how their art is to lead to a Liberal civilization. They must be left free. And it is here that

THE LIBERAL ENVIRONMENT

I fear that in a field which is of great importance we are failing.

The most obvious field of failure, and the one which is most properly political in the sense that it directly affects us in our life as a community, is in the building of towns and villages. The amount of land we have is limited. It cannot be increased to meet extra demand. The layout of our country whether in towns or in the countryside cannot be left to individual profit. Town and Country Planning should be undertaken on behalf of the community. In this context I do not particularly like either the word 'planning' or the word 'community': the first has a repulsively bureaucratic and dead ring about it, the second reeks of social surveys, self-conscious get-togethers and bossy neighbours. But 'Town and Country Planning' is now almost a term of art and the debasers of the words like 'community' and 'social' have not left us with anything to take their place.

The planning and building of towns is a proper job for nations or groups with common traditions. Here the Nation State comes into its own: one of the first things to be done is to devolve power in this field to Scotland and Wales and the English regions. The nation probably must then retain some overall authority but within the nation there should be further devolution to regions and local authorities.

At present building is controlled in several ways: factory building is subject to Board of Trade licensing: there are the Town and Country Planning Acts: there are the by-laws of various cities: but in spite of all the regulations the two most potent diseases of present-day Britain spread steadily. The first of these is the growth of great urban muddles—the spread of London or the Lancashire towns—and the second is the creep of 'subtopia'. The depopulation of the countryside is the reverse side of our unmanageable cities and the wild scramble to get as far away as possible at holiday times is partly the result of urban squalor.

The attempts at green belts, and even the efforts to confine overspill into respectable new towns look like failing. We shall have to take a new look at the whole subject. And when we do so we have to lay down certain things which are most important. First, land in Britain is valuable and must not be wasted. Secondly, communications in Britain must be improved. Thirdly, it is desirable to keep some spread of population throughout the country. But fourthly it is impossible to dragoon people into living in places and in ways they do not like. To start with this last point first: we must beware of assuming that they like what they put up with. I do not believe that the inhabitants of those dun rookeries of mean new tenements you see outside so many towns would choose them if there was any alternative housing which gave them comparable comfort in size of rooms, water, light and dryness. The majority of people in Britain now have to live where they can, and where they can is usually where the local authorities or sometimes a large-scale builder have put down houses. But I think it is probable, and probably distressing, that a lot of people like suburbia. For one thing, they like gardens which in an age in which houses and furniture, motor cars and clothes are all becoming standardized give them some opportunity for individuality, and the creation of beauty, let alone exercise and some vegetables. Far too little attention has been given to the building of good suburbs. Nevertheless, there are many people who would prefer to live in the centre of towns if it were possible.

One way to stop the decay of town centres is through the local taxation system. A local taxation system based on land values would curb the waste of land which now goes on and could be adjusted to take into account changes of taste. Suburbia raises many difficulties: it takes up agricultural land: it involves extra public transport. Yet we continually build new housing estates in the suburban districts and let the houses at fantastically low rents and rates. Glasgow Corporation has let its new houses at an average of under 10s. per week, with gardens and considerable maintenance by the Corporation. They are actually cheaper in some instances than public housing in the Highlands, and very much cheaper, of course, than much private housing on less valuable land. We must reinstate rent as a means of ensuring economy in land and putting it to the best use. The remoter country districts would then regain some of their natural advantages since, their land being less valuable, local taxation of land values and rents would be lower than in or near the cities. But we would also bring pressure on cities to concentrate. Suburban dwellers would pay heavily for their extra space: we might also find that the differential

THE LIBERAL ENVIRONMENT

between values in the centre and values on the outskirts of cities narrowed to such an extent that there was a return to inner districts. Some people will throw up their hands in horror at the suggestion that the cities should be more concentrated. I am sure they are wrong. They are still reacting to the degree of crowding which disfigured Victorian slums. Today the suburban spread is a far greater danger than concentration in well-planned towns where density can be controlled.

Another step which needs to be taken is to stop subsidizing urban transport. The London Transport Board encourages subtopia by shielding those who want to live in the outskirts of London from paying the full cost of their transport to work in the centre. If transport is to be subsidized at all, then it is long distance and rural transport which should be helped.

The location of industry has figured a good deal in the news since the war. The location of schools, universities, ports and other things is equally important. All these, including industry, make a call on public resources in the way of housing, drainage, roads and so on. They can all make or mar a district. A great deal of nonsense has been talked about the siting of industry but in spite of that it is something in which the public have a legitimate interest. The interest must, however, be expressed with care and according to principle. A complete lesson in the wrong way to site industry can be learnt from studying the efforts by rival groups of politicians to get a new steel strip mill built in their areas. Political pressure to get votes or earn prestige is clearly a bad way to plan the layout of industry. Then there is the prevalent assumption that industry exists to give employment. There is the assumption that people should not be required to change their home let alone their job. As regards this last there is a real difficulty for many people, and a creditable attachment to their home districts for others. Liberals should ignore neither of these. But many people actually want to move at least once or twice in their lives. The effort to bring industry to men and not men to industry must be tempered with the thought that many young people want to leave their home districts.

If transport were improved it should be possible to free industry from the compulsion to cluster round ports or over coal-fields, creating markets which are themselves a new draw. Some dispersal to the smaller towns, some combination of industry with agriculture is desirable. There are many small farms which should be run by a family some of whom are employed in local factories. There could well be a financial incentive to go to these smaller towns, partly through the adjustment of local taxation, partly the subsidization of long-distance transport, and partly by a tax on businesses which insist upon setting up in already overcrowded cities. Better trunk roads are certainly essential for any such dispersal of industry. Then it is desirable to help towns which, having been largely dependent on one industry, find that industry failing them. We cannot necessarily expect to keep such towns at their present size. Indeed the decay of an industry may give us the opportunity to replan a town, abolish many slums and make it into a far pleasanter, if smaller, unit. But in some of the cotton towns or in a town like Dundee, the hardship caused by local unemployment can be very serious. I am inclined to think that the development area idea should be extended. Bodies like the Scottish Council for Industry or the Northern Ireland Development Council have done much to bring new industries to hard-hit districts. Industry itself should tackle this problem with the Government by setting up a corporation or corporations with funds and fairly wide powers for the development of areas which are badly hit by unemployment due to changes in techniques or markets or the general pattern of life, unemployment that is likely to be permanent in the particular trades or industries and which is concentrated in areas. The Highlands I would include in any list of such areas.

But the remedy does not necessarily lie in new industry, or new industry alone. There are small towns which would make better homes for universities or technical colleges than do big cities. There are ancillary trades which are hardly industries, tourism, fishing, intensive agriculture on small farms which can all give some small but useful degree not only of employment but of life to remoter districts.

The general outline of Town and Country Planning should be to aim at using our local resources to the best of our ability both from the economic and aesthetic point of view. The overall plan would accentuate the difficulties of crowding into vast semi-urban

THE LIBERAL ENVIRONMENT

conglomerations by making it very expensive—or very expensive relatively to either going out to smaller towns or country districts, or to concentrating in more tightly organized cities. We should embark on a large programme to build better roads and develop air and rail services suitable for these islands. It is, for instance, fantastic to go building faster aircraft needing larger airfields to the exclusion of aircraft with a short take-off and a lower running cost capable of serving our own needs. We should encourage small factories in country towns and even villages employing men and women who retain an interest in the land. And we should take a more positive step to deal with districts with peculiar difficulties by setting up Development Boards with wider powers and better finance.

But on a lower level the work of detailed planning must be done. Here regional boards should be given the widest scope to introduce variety. These boards should coincide with districts which have pronounced characteristics and traditions of their own. And it is here that the artist and the architect should come into their own. And not only the artist and architect, for they themselves are too often centrally trained and imbued with certain negative 'standardized standards'. We have before us the prospect of a world in which every city will look the same. All will be composed of rectangular concrete buildings, some set on their sides, some on their ends. Every village will be a squirming little estate of more or less identical boxes and every public building will be an attempt at modernity without consideration of size, climate or landscape. We must not look, however, to the most thrusting shopkeepers, avid for plate-glass and chromium fittings, nor to the professional arbiters of good taste who too often have no eye and are bemused by 'the canons of good taste'. The best taste is often found among the humbler villagers, farmers, schoolmasters and those who look about them with eyes unclouded either by avarice or theory.

All these should come into their own to experiment. When Liberals say that Town and Country Planning is a legitimate type of planning we are sometimes told that it is the field where petty bureaucracy is at its worst. And so it is now. At present the sifting through committees, the licensing, the by-laws are far more likely to stop a budding Picasso or Epstein or Lloyd Wright than to

THE LIBERAL ENVIRONMENT

save us from a hideous concrete pavement or chromium and plate-glass shop front. But this need not be so. Local Planning Boards should allow the widest scope to individuals and once having put their faith in individual effort should be prepared to take the mistakes with the successes. The kind of board which would do this, which would lay out new houses in proper terraces, for instance, which would build more country cottages of local materials and deliberately exclude traffic from the centres of some older towns would need imagination of a kind that is now rare. But imagination must be found. It must come from a combination of reforms: reforms in education: reforms in local government and the interplay of local and central government and their officials.

No doubt experimentation will be necessary. Take these two possible instances. There are many towns now which have their own societies for the protection of their existing assets in beauty and the provision of new ones. These should be allowed to apply for official recognition. If granted, a local board could be composed of some representatives from the society with some representatives from the local authority. Such a board, if it would take the trouble to find the right officials (not necessarily those who have certain qualifications) might make a much better job of town development than the present local authority. And it should be allowed the maximum freedom from central regulations. Or take an area like Cornwall or Galloway, which still has fairly well marked characteristics. Here, too, a joint board might be much more effective than a number of small building and planning authorities all, in fact, tied to regulations laid down from on high.

What is certain is that Liberals attach the greatest importance to the conditions in which we live: to the use of our limited space, to the duty to preserve what has been left to us and to create more beauty where we can. The location of industry, the development of agriculture, the planning of our towns must all be considered from the humane as well as the economic point of view. Here is a legitimate field, not for centralized bureaucracy but for new experiments by individuals who have something to express and for common action by people who feel a common need and a common tradition. Here is a field in which 'planning' is legitimate and should be purposeful.

THE LIBERAL ENVIRONMENT

Liberals have never paid enough attention to the ways in which a Liberal society gets its information. Television, radio, newspapers, periodicals and books are essential to a good society and to democracy. But they raise all sorts of problems. Genuine problems arise of control, of the reconciliation of freedom with adequate service and of commercial enterprise with true information.

The solution of these problems must lie in having a great variety of papers, periodicals and radio and television programmes. Here, however, we come up against the limitations of demand and in the case of radio and television the restricted number of channels available. Liberals are against any Government control of newspapers or the air. Government subsidies may mean control. But reliance on private enterprise for information has evils of its own. Democracy, let alone Liberalism, could not flourish with wireless given over to entertainment and a public press composed largely of magazines purveying sex, crime, strip cartoons and personal stories. The increase in circulation of the expensive Sunday and some of the expensive daily papers is a hopeful sign that this will never happen. But the Press and ITV depend on advertising and a change in advertising methods might put them in a very unfavourable position. Already many local papers have gone out of circulation or are amalgamated, while lately the demand of big advertisers for either total coverage or a specialized and wealthy public has threatened one or two national papers. Periodicals and 'highbrow' papers are even more vulnerable than dailies.

Liberals by improved education and by a taxation system less deadly to smaller businesses would hope to improve the prospects of papers surviving commercially. As far as radio and television are concerned, new techniques, such as short-wave broadcasting, should make variety more possible. Rather than attempt to impose regulations, Liberals would use the play of assertion, contradiction, free discussion and multiplicity of not only opinion but of reporting as a safeguard for truth. And there is no doubt that if it can be achieved such a free market is the best climate for truth.

But suppose monopoly, distortion, censorship overt or concealed creep in? There are three sides to this question. First, how

THE LIBERAL ENVIRONMENT

to maintain adequate media; secondly, what standards are to be aimed at; thirdly, who is to control the media. The best way at present to give some help to smaller papers without threatening their independence will be through a subsidy on paper and possibly also by the establishment of public bodies empowered to give help on the lines of the University Grants Committee. If there were several of these bodies composed of representatives of regional or national authorities and members from other bodies such as the National Union of Journalists, universities and learned societies with their own funds with which they could help deserving papers, etc., this would minimize direct party political influence and allow some connection between local opinion and the Press. It seems to me that this is another field in which national, in the sense of Scottish or Welsh, feelings and local interest should have a voice. But in the first instance assistance to a free press should come in such general and impartial ways as a subsidy on paper. Any further assistance must depend on circumstances. The central Government should keep as far away as possible from meddling in news or views. This is far from the case just now. It is high time that a check was put upon Government propaganda. Few people realize how far the process has gone. We have in the office of Lord President the chairman of the Conservative Party, one of whose principal functions is party propaganda. Every ministry has its public relations officers. Favours are shown to certain papers, the B.B.C. news seems sometimes biased in favour of Government pronouncements. Worst of all, the heavyweight parties lay down how party broadcasting is to be done and which parties shall do it. They even try to interfere with the choice of broadcasters in non-political broadcasts. These decisions taken behind closed doors and resulting in the total exclusion of Scottish and Welsh Nationalists are quite indefensible. Certainly a Government or any public body must make its decisions known. The Governments of the Western world have a duty to broadcast to people living under Government tyranny—though how they should do it is another matter: an independent body financed by the Government and dealing with the issues objectively is the best method, but the imposition of a closed shop in the interests of orthodoxy and the use of all the machinery and finance of government for party propaganda at home should be stopped.

We often hear of the need for new standards in journalism. No one can doubt that the standards of a great deal of journalism are low. There is the spectacle of men already extremely rich making themselves richer still by turning papers over to pornography and intrusion into private life. But there are also very high standards in journalism. The trouble is that the press owners and editors are as sensitive as debutantes to any criticism. It is seldom that pictures, profiles, or indeed news of its owners appear in many newspapers. With the exception of the *Manchester Guardian*, the *Observer* and occasionally *The Times*, there is practically no criticism of the daily press by the daily press—and a fearful outcry from the subjects of such criticism when it is attempted. One of the most effective and Liberal ways for the Press to set and maintain standards would be for it to review itself: it is encouraging to see the *Spectator* and *New Statesman* doing this.

It is obviously difficult and dangerous—for it may lead to a covert censorship—to lay down standards. But we recognize at least the lowest standards. The intrusion into private lives because, so the Press owners grandiloquently tell us, the public has a 'right to know', is nothing but a sales gimmick and in many cases goes far beyond the bounds of decency. There is no right to pursue people, even public figures, into their bedrooms; on the contrary, everyone has a right to some privacy. Again, if news is presented as news it should be as accurate as possible. If mistakes are made they should be corrected and the corrections given as much publicity as the error. These are elementary rules but by no means universally acted upon. If custom does not enforce this, law will have to.

But when we get to the other end of the scale and look at the good points of the British Press they are very good indeed. Its entertainment value is high. The amount of news in dailies which still aim at being foremost newspapers is high and accurate. The leading articles and special articles of all kinds are excellent. The high standards are there and will be maintained so long as the Press maintains its present shape. But can it keep that shape? Some papers are trying to ensure the preservation of their character by forming trusts such as the trusts which control *The Times*,

THE LIBERAL ENVIRONMENT

Manchester Guardian, Observer and *New Statesman*. This seems to be a development which Liberals should welcome. As far as I know, the Press Council could not at present concern itself with a change of ownership in newspapers which might lead to a falling off in standards. But Liberals want a Press Council on which the public is represented and which is prepared to take a more positive line than the present body. If this came into being it might at least make some comment on general as well as particular lapses on the part of newspapers.

The formation of trusts may do something not only to preserve the standards of certain papers but to prevent a monopoly of ownership. But trusts are as liable as any other form of ownership to become frozen. There is a Conservatism of the Left as well as the Right. With all their virtues they may prevent the papers they control changing position so as to remain relatively in the same place. There must in a Liberal society be plenty of room for nonconformists and they must have means of expression. But today it is very difficult to found a new paper, weeklies and monthlies languish, where will the iconoclast of twenty years hence find expression? The answer is that he will probably find it somewhere, somehow, if he really has anything to say. But that answer is not good enough. The need of the mass circulation Press to offer what the mass advertisers want, the squeezing out of smaller papers and the difficulties of some weeklies could mean that certain individuals and interests established a stranglehold on news and comment. As I have suggested, it might be possible to avert this by setting up a body as independent as possible to give financial help to threatened papers. But it is difficult to see such a body giving help on a scale sufficient to keep a national daily in existence, except temporarily, without interfering with the freedom of the Press. It might effectively help the *New Statesman* or *Time and Tide* if they were to get into difficulties or it might help specialist papers or local papers if any of these needed help—and it might be a useful service—but it would need very large funds to finance a daily and by using them it would affect the sales of other dailies. Further, it would not be an ideal body to finance revolutionary publishing.

The most hopeful experiments might be made over the air. Here

THE LIBERAL ENVIRONMENT

we have a new medium of great potency with no set customs. So far the B.B.C. and the I.T.A. have done a reasonable job. Many of their faults are imposed on them, their virtues are largely their own. But the people who are appointed to their controlling bodies are of the official orthodoxies. These appointments should not be left in the hands of the Government. Broadcasting and Television should be decentralized. Appointments could then be made by national Parliaments or regional bodies and even by direct election. Control of Television is a matter where direct election might have its uses. This is not to say that some services should not give coverage to all the British Isles, it is a plea for variety, for local and specialized as well as national services. Further, universities should use the air as they do in America, and I should hope that other bodies might follow them. The British Association, the Saltire Society, the city companies and some of the theatre companies, to mention four sources only, might produce programmes occasionally if not regularly.

I am not against the financing of Television by commercial advertising, but it should not be the only method. We should supply 'Third Programmes' on the air even at a loss, even to a small public. Such programmes should not be manned solely by orthodox highbrows. Specialist programmes, programmes which shake the conventional Left and Right, educational programmes must all have their place. No one can devise a place for revolutionaries, its creation would stop the revolution, it would be an altar to an unknown god. But while on the one hand the general means of spreading opinion and information should broadcast the best that the orthodoxies can give, it should be loosely enough organized to let the unorthodox in. It can be done by ensuring variety and by keeping total control out of the hands of commercial or political or religious interests.

The disease of rigidity, the stifling effects of orthodoxy can be seen throughout our society. No Liberal can be complacent about many of our social and legal customs.

Patronage and privilege are still rife. We have done something to get rid of the old class system: if we have not got rid of it entirely we have at least fragmented it: but it still has a grip on political and social life. Discrepancies in wealth will always give rise to

THE LIBERAL ENVIRONMENT

some distinctions, but the United States of America has shown that these need not be serious. The crux is that the field should be open. It is the tie-up with hereditary advantages and political patronage which is damaging. Certainly there is a case for limiting inherited wealth. This is one reason why Liberals prefer, to death duties on the estate, legacy duties where the incidence of tax can be varied with the amount left to the individual legatees, so that the diffusion of wealth by spreading it among many beneficiaries is less heavily taxed than the same amount to one person.

The system by which all sorts of new jobs and opportunities are made available by the ruling political party is wide open to criticism. In a pluralist society there is no reason why the judiciary and the ecclesiastical authorities should not make legal and ecclesiastical appointments. To suggest this is not to imply that jobbery has been rife. It is, however, to protest against the concentration of power in the hands of the Government. I know that honours and awards are greatly valued. They are a way of rewarding people who have received inadequate recognition in other ways. Unquestionably they have their place. But the Honours List is inordinately long, and many of the selections are either routine appointments or hardly seem to justify their presence. A Liberal Society certainly must not be a dull and undifferentiated society. But, equally, those whom it honours must be genuinely meritorious. We are a very long way from the ideal state of affairs. While the Prime Minister must keep his pre-eminence as the Crown's adviser, I see no reason why in all the spheres outside politics the Crown should not seek advice more readily from other sources. The Privy Council, too, might assume responsibility for many matters connected with patronage and with the arts.

It is more important still that many of our laws should be in keeping with a Liberal society. Cruelty to children is still treated as a lighter offence than crimes against property. Sexual offenders, especially homosexuals, are treated too much as criminals and too little as invalids. So far no action has been taken on the Wolfenden Report. Betting and licensing laws need amendment. From our lunatic asylums come disturbing rumours of cruelty.

Civilization must be held to account for its general outlook on all these matters. In this field, as in politics, as strictly defined today, we have failed to keep up with the best that reason can teach us. There is little pressure behind reform of this kind and so it is neglected. This neglect is another sign of the need for Liberal reform in our society and in our political organization.

9

British Liberalism and the Rest of the World

One aim of Liberal foreign policy is to spread Liberal ideas. Liberalism is the enemy of Communism and it cannot stand on the defensive. I am not clear where Western Conservatives or Socialists stand when we face the issue of defeating Communism. They sometimes talk about the battle for men's minds. Their Governments broadcast propaganda. Yet from time to time they disclaim all intention of interfering in the internal affairs of Communist countries. They frequently denounce subversion. Yet if propaganda is effective it must be a method of interfering in the internal affairs of countries behind the Iron Curtain. And the interference must have as its ultimate goal the 'subversion' of Communist Governments. Liberals mean to subvert the Communist Governments, though they do not mean to do so by war, assassination or similar means. The dilemma shows up the futility of non-Liberal opponents of Communism. As a result of their failure to think out what they hope to do, Western counter-offensive to the offensive of Communist ideas is ineffective. It fails very often to treat individuals as though they had minds, or to appreciate that those minds are battlefields on which our tactics have to conform to the layout of the ground. We have failed to show how democrats would deal with the problems which Communist countries present.

The Conservative would probably say that our propaganda should be largely negative in character, exposing the villainies of Communism: he might add that in so far as it was positive it should show the virtues and superiority of Britain and of democracy. And I suppose that it is on the theme of democracy that the Socialist would play hardest. For Socialism is hardly something about which Socialists can claim to know more than Communists. When preaching to Communists Western Socialists must lay the emphasis on the possibility of going a certain way towards Socialism without abandoning parliamentary democracy. But since 1945, when it was claimed by British Socialists that they would get on better with the Russians than anyone else, we have not heard much of a specifically Socialist foreign policy.

Liberals do not believe that it is enough to counter Communist lies: they do not believe that it is enough to point out the advantages of living in Britain, nor even to preach the merits of democracy. Some positive alternative to Communism is needed. The alternative is individual liberty and opportunity and the respect for each other's welfare and freedom which must go with it. This can have an appeal, if properly handled, to everyone. It does not require the acceptance of the British way of life, nor of parliamentary democracy as we practise it.

But that the preaching of Liberalism has its dangers, I would not deny. It may conflict with another aim of a Liberal foreign policy, which Liberals share with Tories and Socialists, and that is peace. Again, Tories and Socialists have not defined the purposes for which they are prepared to go to war. I do not blame them, nor do I suppose most Liberals are any clearer on this topic. Most of us who are not pacifists will in fact fight when our Government tells us to do so and we should take it for granted that our Government would not declare war on a scale likely to involve us in world war if some vital British interest was not threatened. But 'vital British interests' in today's world are very different from what they once were. Some of those who are in favour of unilateral nuclear disarmament would say that since to avoid annihilation must be a vital British interest, and since annihilation is certain in a world war, no British Government should ever commit us to such a fate. For them the avoidance of nuclear war in any circumstances is the only vital British interest. Others would say that to strengthen international law and order is of such vital importance to Britain that we should only go to war in a cause approved by the United Nations. The nuclear bomb and the United Nations are only two of the new developments which confuse still further the never simple idea of vital British interests.

We can say that the maintenance of peace between States has a big claim upon us. Liberals would go a long way to maintain peace. But there are clearly other claims to be met. The defence of Liberalism has a claim on us. Ultimately, we are not prepared to accept tyranny, the extinction of Liberal institutions and the suppression of Liberal values for ourselves and our friends and fellow-countrymen without a fight. But what do we say about the suppression of Liberalism in other countries? This is a difficulty—and a very real one. Is it our business to fight for any country threatened with external aggression if the aggressor is illiberal and the threatened country liberal? I do not believe that there is a satisfactory theoretical answer to this applicable to all cases. The most we can say, I think, is that such a country in such a situation has a claim on us. There will also be a stronger claim on us if the United Nations has called on us to take action. But the decision, to go to war or not, must depend on how strong the competing claims are in any given set of circumstances. What Liberals must admit is that there is a claim on us to fight for Liberalism. There was such a claim to go to the assistance of the Hungarian uprising. That our intervention might have been ineffective and might have led to greater disasters than the suppression of the rising set up stronger claims for non-intervention. But we cannot give peace absolute priority. We cannot give the United Nations absolute priority, nor the decisions of our own Government. The decisions of these authorities are factors to be taken into consideration—they create further claims on us. But Peace or War is ultimately a matter of judgment and a war is not made right or wrong simply because, say, the United Nations has declared for or against it. The risk of precipitating war is not a reason for ceasing to preach Liberalism, even behind the Iron Curtain where the result may be an uprising involving us all. But I must emphasize that for Liberals force is not the only weapon. Certainly the mere fact that one more or less Liberal country attacks another, or one more or less Liberal Party within a country carries out a *coup d'état* is not a reason for armed intervention. Liberals do not feel that there is any virtue in maintaining things as they are—if their present state is unsatisfactory. They are not prepared to go to the aid of any Government endangered by 'subversion' merely because it is the established Government.

BRITISH LIBERALISM AND REST OF WORLD

This brings me to the third aim of Liberal policy—the creation of a means of peaceful change and an instrument to carry it out. This may well be said to be the central problem of all politics, one which few countries have solved internally and which we have hardly begun to tackle internationally. Just as the championship of Liberal ideas even when they may lead to revolution may in some circumstances conflict with the aim of peace, so change may conflict with the defence of what are considered British interests. Liberals, so long at least as the nation-state wields anything like its present power, acknowledge that it has a responsibility to look after its citizens and its citizens have an obligation to support it in many circumstances. But this responsibility will change as supra-national organizations take over some of the sovereignty of the nation-state and, in any case, today the interests of British citizens are certainly not best defended by gunboats or by attempts to freeze the present position. There is, for instance, no reason to suppose that it is in the interest of our people to cling on to colonies for ever—we have seen the benefits of surrendering to change in India, we have seen the great results of the Durham Report in Canada and in many other parts of the world change has strengthened not weakened British influence. To defend British interests certainly is an aim of Liberal foreign policy but British interests must be interpreted in the light of the present situation.

Among British interests trade is one of the most important. Trade and the development of the world's economy, the improvement of life in the poorer countries, the maintenance of prosperity are indeed both a British interest and in their own right an aim of Liberal foreign policy.

In rough outline then, the aims of Liberal foreign policy would be to promote a Liberal world, by peaceful means so far as that may be possible; to make it more possible by helping to create international authorities through which peaceful change could be agreed and enforced and all the time to work for freer trade and greater prosperity with due regard to the true interests of the British people.

As to the methods of policy, this subject can be divided into two parts, these parts being divided by the great division between the Communist and non-Communist countries. There is the question of our relationship with other countries and peoples outside the Communist world and the question of our relationship with that world. The latter question has obscured the former so that we sometimes speak as though if only Russia and China were removed from the planet or converted from Communism all our troubles would be at an end. But this is not so. Apart from the difficulties which might still arise with these large empires outside the European tradition, even if they were not Communist, there are very many problems of development within the free world itself as it at present exists.

As far as our policy towards the Communist powers is concerned we must be prepared to alter it as changes in their points of view or those of our allies make it necessary to employ new methods to reach our objectives. One of the most serious weaknesses from which the Western world suffers in its dealings with Russia is that in addition to lacking any clear, simple idea of what it wants it lacks the centralized directorate which could pursue it. But granted that we may have to deviate often and rapidly from the path, what path should we pursue towards the object of achieving a more Liberal and a more peaceful Communist *bloc*?

Though, as I have said, I do not think that we can rule out an ultimate resort to war, Liberals must start on the assumption that we are going to negotiate with the Communist Governments and argue with their citizens. We aim to convince and not to frighten. This may seem obvious. But there are well-informed and liberal Americans and, indeed, many other observers, who think that it is not possible to negotiate with the Russians. They would say that it has been tried again and again with results which have been on balance far more harmful than good. They would point to the very nature of the present Governments of Russia and China and to their refusal to take up Western offers on disarmament or to honour the spirit in which the West made considerable sacrifices at Yalta. But there are stronger reasons for trying to negotiate with patience; though we should not expect too much, especially from short, high-level meetings. The Russians have not always broken their commitments. Austria has at length been settled. The art of negotiation is to choose matters from the settlement of which both sides have something to gain: and also to realize that negotiations must be genuine—both sides must be willing to give way in some degree. There are from time to time subjects upon which it at least seems to be worth opening negotiations: at present disarmament in the Middle East and military disengagement in Central Europe seem to be such subjects. Such negotiations may fail. The West must be strong enough to break them off for a time and resume them again in some other field. As in their inevitable differences among themselves, so in their need to report successes the democracies are at a disadvantage. It is difficult for a Prime Minister dependent on free votes to come back from a diplomatic meeting and acknowledge utter failure. But if that is an argument against pinning too many hopes—and too much publicity—on summit meetings it is not an argument against any negotiation whatever.

The Prime Minister recently (20th December 1957) said in the House of Commons that our policy was to support 'a firm and powerful N.A.T.O. from the military point of view, but always ready to discuss and to negotiate on a practical basis to obtain practical results'. That is all right as far as it goes. But it seems to Liberals that we have not in fact over recent years taken the initiative with sufficient zeal in negotiation and, while practical basis and practical results are what we all want, successful negotiation needs a clear-sightedness and a certain willingness to seize opportunities, even to take risks, which have been lacking. We have, for instance, let our opportunities for a settlement in the Middle East slip by for lack of any policy other than vaguely supporting the dynasties in Jordan, Iraq and the Gulf States and giving a half-hearted guarantee to Israel. Military disengagement but full diplomatic engagement, leaning up against the East in continual negotiation, should be our method.

With the policy of negotiating over outstanding causes of friction with Communist Governments should go a campaign for getting at the minds of men behind the Communist curtain and in the 'uncommitted' countries. But there must be a theme. There must be faith in the West in a way of life which offers an alternative to Communism. In Liberal eyes this must be Liberalism. Certainly it cannot simply be the boosting of Britain or America or

France or any other country. For in the modern world as well as the foreign affairs which are carried on through embassies and between Governments, there are the foreign affairs carried on by a direct appeal from Governments to the subjects of other Governments. If this is subversion, it is no good trying to denounce it. On the contrary, it gives the West its only chance of changing the nature of the Communist régimes and thereby removing one potent cause of war.

The propaganda of Liberals must be partly negative. It is vital to remove the barriers which have been thrown up in the last fifty years. Let Trade and Tourism flourish: reduce tariffs, quotas, restrictions on travel. Anyone truly rooted in the spirit of Liberalism will be a more potent agent than a Communist. Apart from this, prosperity and happiness are the allies of Liberalism. But there must also be a more aggressive side in the presentation of Liberal ideas. The Russian, the Pole, the Siamese or the Iraqui are not going to be won over by blatant propaganda designed to boost the West as an alternative master to Communism. They are to some extent uninterested in the struggle of the free countries against Communism. They want to improve their own lives. They know that they have their own problems which differ from those of either Western or Communist countries. They want help politically and economically—but help to enable them to help themselves, not help merely as a bribe to become an inferior ally. We must show them that it is not a question of asking them to stick to our side—but our business is to convince them that Liberal values and Liberal methods are as valid for them as for us. It is also a question of showing them that we in the West are the dynamic societies with the answers to modern problems. The most pressing of these problems is the political organization of the free world.

Already the boundaries between sovereign nation states have become dangerous anachronisms. Industry is international. The great firms of modern business operate all over the free world. Finance is international. No country can save itself by itself from slump and booms even though Governments still sometimes pretend that they can control their own economic climate. No country can defend itself. No country can carry on its own foreign or colonial policy without caring what the rest of the world thinks.

If it were demonstrably better that the world should be divided into sovereign states, Liberals might not be daunted by the appalling task of keeping such states in full independence. But every Liberal argument points the other way. Every Liberal must rejoice that developments in politics and science point to unity not division among the people of the world. If we are to have peace it is essential to break up the bellicose robber baron states who have torn the world apart in the last fifty years. Peace will only be guaranteed by a world authority, and even if that is a long way off, every step towards it is welcomed. If we are to raise the standard of living of our own people, let alone the people of Asia and Africa, that can only be done by allowing men, capital and goods to move freely and peaceably. If we are to meet the problems of population or of education, or of political organization in Africa, it needs an effort transcending national boundaries. Indeed, since at the root of Liberalism lies the belief in equal opportunities and rights for all men, the unity of mankind must always be stressed by Liberals. It is our opponents, therefore, who are faced with the gigantic task of damming back all the forces making for unity. Liberals are rowing with the tide of change. But change may be violent or peaceful. It may come about by Liberal or Communist means. This is the problem of what are called in old-world language 'foreign affairs'. That home and foreign affairs are inextricably mixed up is reflected in the loss of prestige of the Foreign Offices. Only a Foreign Secretary of quite exceptional character can make a policy at all. Yet we cling obstinately and harmfully to the picture of a bygone diplomacy: a diplomacy which acts as though the units of the world were still the nineteenth-century independent states.

There was a time when Liberalism was naturally associated with the rise of the independent states in Europe. For in Italy, or in eastern Europe, the rise of the State was a protest against oppression. The State was a means of gaining political freedom, and even economic prosperity. But even then Liberals were firmly opposed to militarism, to the raising of trade barriers, to the virulent national Socialism which reached its peak under Hitler and Mussolini. It is the combination of Statism or Socialism with nationalism which has led to warmongering states from which we have suffered such calamities in this century.

It so happens that few peoples have so much to gain from greater unity in the world than the British. One aim of Liberal foreign policy must therefore, as I have said, be to make a peaceful and liberal transition to new political organizations. Without such new political organizations we are neither going to remove the causes of war nor create an authority capable of enforcing any sort of international order and peace.

But though we live at the tail end of the age of the nation state the nineteenth-century sovereign state is still very influential. It bounds our horizons. We are loyal to it, our political hopes are its hopes, our careers depend upon it. More and more people are dimly and uneasily aware that it is an out-of-date idol, but the more uncertain some of its worshippers feel, the blinder their devotion to it.

The state of a man's affections is a fact. And that he will have affections is an even more obvious fact than the fact that nationalism is a very strong affection. But loyalty to his fellow-countrymen or pride in his traditions is not the same as obedience to the comparatively new and passing phenomenon, the nation state. Men are not, for a very long time at least, going to be governed by grey impartiality; they will like one country, one place, one race more than others. They will need some mystical body in which they can lose themselves. They hunger for causes over which they can wax enthusiastic, and symbols which typify those causes. Nothing of this is necessarily illiberal, nothing is necessarily evil. For many purposes the nation state will be a useful unit. But not for all. It is its claim to absolute loyalty and it is the incompatible absolute claims of all states which are evil and illiberal. It is the attempt to squeeze everything within the bounds of the nation state and build it up in opposition to the world at large which may lead us to destruction. Liberals repudiate these pretensions of the nation state. They do not merely want to limit them. All phrases such as 'interdependence', 'partnership', 'international co-operation' are used today to describe relations between states. They are variations of the old theme of alliance. They may be useful as far as they go, but they do not go far enough. The only partnership which has meaning today is a partnership between people, not states, the word 'supra-national', ugly as it may be, is much more hopeful than 'international'.

But we must engage men's emotions in new patriotisms, not bounded to the old bellicose, xenophobic nation-state symbols. Here we can learn something from Communism. Historically nationalism and religion have been the main driving forces of mankind. Communism has combined them into a synthesis, and its successes and failures should be carefully considered.

It is frequently said, with some justification, that political Communism is a religion. It may be in some senses. It is equipped with many of the trappings of religion. It has its orthodoxies and its prophets. In its glorification of inanimate objects and its repudiation of the individual, it is, in Russia, rather like a primitive religion. But Marxism is not a religion and books on Communism are quite different from religious books. Its appeal is not wholly religious. Its appeal lies in its ability to offer an explanation and a code of conduct. It offers a simple and not wholly illogical answer to many questions which puzzle mankind. It can tell the ordinary man what he ought to do. It has the simplicity of great movements: the simplicity which Western democracy lacks. The West, if faced with the need to do anything, does as little as possible, stressing the complexity of the problem and ravelling up the skein still further. Can you increase trade in Europe by the simple expedient of removing the tariffs which impede it? Not at all. You must have prolonged discussions and invent new ladders to get over hurdles of Europe's own creation rather than removing them. The Western world takes refuge in complication. Having hopelessly exacerbated Turk and Greek Cypriots, Britain at last propounds a constitution of almost incredible complication, rather than cut some of the knots of her own tying. Communism, too, has succeeded in involving ordinary people. It is a paradox that in the democracies which claim to be run by their citizens, these very citizens are becoming more and more detached from political action while political Communism, authoritarian though it is, in its proselytizing stage at any rate, succeeds in calling out real individual effort. It is paradoxical, too, that Communism which is a doctrine of the *élite* has got across the simple idea that it is on the side of the poor. In its relation to nationalism, Communism has succeeded where nationalism has been largely a protest against the successful foreigner. It has succeeded where its rivals have behaved as colonial powers. Its success has been partly fortuitous: had the first Communist country been a colonial power, or had its satellites been separated from it, it would not have been so easy for it to play on anti-colonialism, but the Russians being themselves until recently under-dogs rubbed no coats the wrong way and raised few hackles in their dealings with other under-dogs. This spirit of comradeship amid under-doggery may continue for a long time as emergent countries still look to Russia to see what can be done with a start from scratch (as the Russians have convinced the world is the case, though Capitalism, in fact, did a good deal for them) and without Western aid. The Russian Communists have not discriminated against black races. They seem genuinely free from the snobbishness, the air of superiority, the Herrenvolk attitude which has done the West far more damage in the eyes of Asia and Africa than its exploitation or political oppression.

On the other side of the balance, however, it must be remembered that the Russians have had little success in their dealings with adult patriotism in eastern Europe, and none with the positive and proud inheritance of Jewry.

The experience of Communism reinforces the obvious world policy of Liberalism. Its appeal must be to the people, to the individual. Its appeal must be simple. And it must make people feel that its fight is their fight. It can then engage the emotions.

The failure of Britain since the war in areas like the Middle East has been a failure to speak to people and win their personal sympathies. We have tried to deal with Arab states in terms of Britain and the policies, desires and objectives of the British State. But the Russians and Egyptians do not try to impress a Russian or Egyptian point of view on other Governments; they appeal over the head of the Government to the people themselves and they do it in terms of nationalism or freedom.

No doubt one of the most powerful motives in the world is nationalism. To a Liberal it is by no means an ignoble one. It is only dangerous and self-defeating when associated with absolute sovereignty of the nation state. In many parts of the world nationalism crosses state boundaries: Liberals should encourage it to do so. The British are guilty of bringing up the peoples of Africa and Asia in the outworn ideas of fifty years ago. Mr. Macmillan, on the 20th February, said in the House of Commons that the tremendous acceleration of nationalist aspirations 'was welcome for it has indeed been the deliberate British policy for over a century to encourage these movements'. But unfortunately we have too often encouraged them to be narrow, selfish and obsessed with the need to ape European institutions often unsuited to other conditions. Incidentally, Mr. Macmillan's words—all too true—reflect oddly on the intense opposition of violent British nationalists to Colonel Nasser and his violent Egyptian nationalists. British nationalists have led African nationalists to suppose that nations can be supremely independent. It is now a myth. In so far as nationalism leads to variety in outlook, philosophies and art, Liberals must welcome it. They must fight against the creeping paralysis of centralization and sameness. In fact, so far as nationalism is a liberating and not a constricting movement Liberals are for it. They are, for instance, in favour of the Scots being allowed a large measure of self-government and encouraged to develop their own tradition, but they are not in favour of separation of Scotland from England or the Commonwealth.

But in the state of the world it is not enough merely to state that over many activities, if we are to be truly free, the State must relinquish some of its sovereignty. We need some wider symbol which may engage our sentiments as well as our reason. A symbol which is not too abstract. A symbol within our grasp. Europe can be made such a symbol for some purposes to Europeans; there are others emerging in Africa and Asia. Instead of shuffling along pointing out the difficulties in such ideas we should be at the head of these movements proclaiming that we and other Europeans are Europeans in action and that the people of Ghana must look beyond the boundaries of their new State. The Commonwealth could be made just such a symbol for many of its people. But a symbol to be effective must be more than a name or an idea: it must be brought down to earth in institutions and in institutions which work. British Governments of late have been very frightened of institutions. They pride themselves on informality, on the genius of Britain and her Commonwealth living in the family atmosphere; they take pleasure in the haziness of the whole Commonwealth idea, no decisions are ever reached or enforced. They have carried this attitude over into their dealings with America and Europe, and they have become adepts at propounding difficulties in the way of any closer co-operation. But in some directions progress can only be made by creating appropriate institutions. As institutions take root and are seen at work, they generate loyalties and in the long run the survival of Western Europe or the Commonwealth or N.A.T.O. depends upon the growth of a common loyalty. It is a mistake to think that institutions do not matter, it is a mistake to believe that a common outlook is enough without any institutions to make it effective. The two must go together. One of Britain's great contributions to political life has been institutional. It may be that our fear of experimenting with institutions is a sign of political decay. The institution will be successful if it can win loyalties and it will win loyalty if it works. But the word 'institution' sounds rather grand and complicated. Institutions should be simple. They can be flexible. They can be mere meetings in private. But they must exist.

The British Commonwealth has a great opportunity for leading the world with the development of political institutions so long as it looks beyond the state of political development, which it, itself, happens to have reached. The lesson which the British have to give to the world is a lesson in how to carry out political change without violence. This has been done by changing the function, if not always the name, of our institutions.

At present the Commonwealth and Europe are the only international ideas which offer a symbol competing in its attraction with the nation-state. And within the Commonwealth the strength of the Commonwealth idea varies very much. Individuals within the Commonwealth feel some warmth towards it, a warmth which goes further than merely approving the idea and wanting their country to be a member of it. The Commonwealth has great potentialities for promoting Liberal values and removing the causes of war. But this potentiality does not lie in its present power as a world force of the old type. If it is going to compete as a sovereign state with Russia and America, it is clearly weaker and in any event the effort will probably kill it. The Commonwealth potential lies in its being a new form of political organization. The citizen of New Zealand or Pakistan can enjoy the responsibilities and privileges of being a New Zealander or Pakistani, but he may also, if we are skilful, find an outlet on a wider or different field. He is like a medieval clerk in Holy Orders who for some purposes owed allegiance and received protection from his feudal lord in whose service he might also advance—but at the same time his prospects were not limited to the feudal sovereignty in which he was born; through the Church he might reach greater opportunities. It should be an object of British 'foreign' policy as much as 'Commonwealth' policy to build up Commonwealth services. Individuals from all Commonwealth centres should be employed in a service available throughout the dominions and colonies. By that means and through the development of Commonwealth organizations of all sorts we could show what living in a Liberal world would mean.

Europe, too, arouses some feelings of affection and familiarity in the citizens of western European countries. Here, too, Liberals should not be frightened of expressing this feeling in institutions. Conservative and Labour leaders have been enthusiastic about European unity when out of office. They have not been so eager for it when in power. Even the man who saved Europe, Sir Winston Churchill, later did her disservice by raising hopes so high by his championship of the cause of unity only to turn a cold shoulder to it when he was Prime Minister. Neither Conservative nor Labour Governments have done more than accept invitations to take part in the various consultative bodies. They have shown no desire to take the lead towards closer co-operation and have been opposed to allowing any executive power to European organs. Liberals dissented from the original decision not to take part in the Iron and Steel Community. A Liberal foreign policy towards Europe would be based on the firm belief that Britain is a part— a leading part—of Europe, and that international bodies should be executive and not merely advisory.

The third grouping which arouses some response from the heart as well as the head is the Atlantic Alliance. Though the ordinary man's attitude towards the U.S.A. is a love-hate one in which jealousy and irritation are often exacerbated by a common tradition which makes disagreements all the more exasperating, it has an emotional side to it, and on the whole there is a greater common feeling than mutual antipathy. This is a field in which the harsh needs of defence seen against America's predominant strength, have led to some common executive organizations and a considerable surrender of sovereignty in fact if not in theory.

The opponents of those who want to look at foreign affairs from a wider standpoint than the nation-state sometimes object that if any attempt is made to press international action too far, if executive and not merely advisory bodies are set up and States yield some real power, irreconcilable conflicts will arise. I do not share these fears. Much progress in politics has come by the narrower institution, the tribe or the feudal baron, surrendering some of its power to wider combinations. Now that it is coming to be realized that a plural society in a country has many advantages, we should consider how far the same sort of arguments can be used in support of a plural international society.

First of all, we should rid our minds of the belief that there is in fact now a great clash of interests between these various groupings. It is not true, for instance, that our Dominions are opposed to Britain's entrance into the Free Trade Area. Secondly, we must accept that institutions will ultimately be judged by their utility. The utility of a Commonwealth Economic Council would be considerable for there is a common interest in many economic matters throughout the Commonwealth. But it is probable that it would need the closest collaboration with America to which the Commonwealth looks for capital. In Europe collaboration is already taking the path of combination for certain practical purposes, power, iron and steel, atomic development. This is the best way in which to begin. Within N.A.T.O. again it is a practical purpose, defence, which is forcing the nations together. A pluralist international society need not fear clashes of interest too much. But from any consideration of the ways in which collaboration is actually coming about it appears that the three groupings of the Commonwealth, Europe and the Atlantic Alliance have many purposes in common. It would be rational if they in turn collaborated more closely. For instance, Australia looks to America for help with both capital and defence. So does Europe. Along the line of developing a common loyalty and common institutions for members of all these groupings would seem to lie a fruitful road for a Liberal foreign policy.

The U.N. group of international bodies inspires little direct affection. Support for the United Nations and its agencies is largely intellectual. Though any Liberal must be a convinced supporter of the United Nations, it seems to me that some Liberals have been too prone to concentrate their efforts on supporting the U.N. to the exclusion of other forms of collaboration. The United Nations will itself rest on a firmer foundation when international collaboration between races and groups of individuals who feel alike has become more normal. When nations are willing to give some executive power to regional bodies there will be more chance that they will make real surrender of their rights to the U.N.

But this does not, of course, excuse Liberals from lending their efforts towards improving the United Nations. There has lately been a stronger demand for some United Nations force. Liberals have led this demand. They see great merit in such a force, even if it were small. Its function would not at first be to challenge national armies. It would come between combatants to be shot at rather than to shoot: to observe and report rather than to enforce its will. Such forces have already proved useful: the U.N. forces on the Canal have been treated rather as police while those on the boundaries of Israel or the Lebanon are described as observers: but the same type of force can perform both functions. It is the presence of an international force which is the tangible expression of the dominant will of the world which matters. Few countries will care to resist such a force even if militarily they could easily overcome it.

It is a valid point that any such force will in the meantime be of doubtful use in a situation involving the territory of the present Great Powers. That is not, however, a reason for believing that such a force will be useless, still less is it a reason for doing nothing to increase its chances of success. A more serious difficulty arises over the control of the force. Indeed, unless states make some surrender of sovereignty this difficulty is in the last resort absolute. A police force needs a legislature to change the law when necessary and an executive to give instructions when the force is to act. The two need not be precise, nor precisely differentiated, but there must be some means of changing the law and some means of calling the police rapidly to meet any criminal activity. At present the United Nations do not have any means to decide on peaceful change. It is not the machinery that is lacking but the power to work it. There is little chance that any country would, for instance, cede part of its territory with any good grace merely because the U.N. decided that it should do so. Suppose that the growth of the populations of China or Japan raises a demand for more living space such as we have had in the past. Can we feel any confidence that the United Nations could handle the problem? I do not. U.N. was much more successful at the time of Suez than many people dared to hope. But Suez was a crisis. A crisis in which no nation was prepared to risk a world war and in which the culprits were two of the satisfied democratic powers. The Suez war does not prove very much. The United Nations had failed to exorcise Arab-Jewish hostility in the years before Suez and it might very well have failed to stop the Israelis before they reached Cairo had not the situation been entirely changed by Anglo-French intervention against Egypt. From the history of Suez, Lebanon and Jordan it may look as though a new method of handling international crises was being developed. Let them come to the boil, then let the democracies intervene and hand over to the U.N. a stabilized position which both can and must be discussed. I believe it would be very dangerous to rely on such a pattern of events being dignified into any sort of system. What is required is some means of resolving international trouble before it reaches boiling point. In some cases this can best be done by regional bodies who have achieved something more like supra-national authority. States which would not regard a recommendation of the U.N. as anything but an insult might listen with a little more goodwill to recommendations of their chosen allies. The framework of the Brussels Treaty, for instance, might be expanded to deal with the problems arising in Europe.

In Articles 39 to 43 of the U.N. Charter itself are set out the various steps which may be taken by the Security Council after it 'shall determine the existence of any threat to the peace, breach of the peace or act of aggression'. There is also now the 'uniting for peace' procedure. U.N. has been invoked several times under these procedures since the war, in Spain, Greece, Indonesia, Israel, Korea and Suez. The present procedures differ in two significant ways from those embodied in the Covenant of the League. The emphasis now is on the U.N. calling on individual States to take action 'by air, sea or land forces as may be necessary to restore international peace and security'. These provisions in Article 42 go beyond the Covenant and unlike the Covenant there is in the Charter no specific obligation on member nations to take action on their own as soon as aggression takes place. The other difference is that under Article 43, though the international force is to be composed of national contingents and is to be raised for specific purposes only, it is to be under U.N. direction. It is indeed to be under U.N. command. Liberals should work to open further these doors leading to international co-operation for peace. It may be that to make these articles more effective, the Security Council may have to cede some of its authority to the Assembly. But while not ruling this out, I should prefer to see further efforts made to work through the Council. For the Council represents the reality of world power. Further, it is a more suitable executive body. And if it is argued that the veto will always make it ineffective in any really serious situation, my answer would be that there is a great deal to be said for continuing its comparative successes, for dealing with not so serious situations before we push it too far.

The Russians already co-operate to some extent and this is to be encouraged.

For co-operation in practical work is the real hope and test of the U.N., U.N.E.S.C.O., W.H.O., U.P.U., I.T.U., W.M.O., I.A.E.A., which may not mean much to us, but they are organizations in which East and West collaborate. And in addition there is the World Bank and the Fund, F.A.O., I.C.A.O., G.A.T.T., which are of growing importance in the free half of the world.

That there are great difficulties in international political development—and in the change from international to supranational development—I cannot deny. A pluralist world is going to raise problems, though not as great as those it solves.

One obvious difficulty is that if government is to be conducted by counting heads there are a great many more heads in Asia than in Europe. What weight is to be given to population? The prospect of politicians promising the starving millions of India that they will be fed and clothed by a redistribution of European wealth seems too near if Indian and Britain were to sink some of their sovereignty together. Liberal comment on this kind of prophecy must be to point out that it is not intended that the nation-state should disappear—nor that it should be shorn of its powers at one stroke before any other institution is ready to assume them. No Liberal either is committed to the surrender of his political rights to a wholly illiberal political organization. We aim at a Liberal world. If Liberal values are accepted in India, then I certainly welcome close collaboration with Indians. Though, as I have pointed out elsewhere, I believe Liberals and indeed all democrats have a lot to learn about the pressures generated in any political organization, we have much to do in developing new institutions. We have much to teach about the assumptions of sound government. All this applies in Asia as much as Europe and it applies to the creation of international bodies. But the starting point must be the growth of greater cohesion among peoples who are already close together—as are the Europeans and the members of the Atlantic community. Further, we must get away from the idea that the counting of heads is the only form of democracy.

Without departing from the ultimate end of achieving the widest unity, especially among free peoples of the world, Liberals must stress that such unity is likely to be stronger if it grows up from those areas which already have considerable homogeneity. Indeed, there is a danger that by attempting to spread the umbrella of international unity too widely in its early stages we shall weaken it. Western Europe seems to be eminently an area in which we could press strongly for unity at once. When it comes to devising institutions which include the peoples of the world, very much more thought needs to be given to the development of new management techniques for such institutions and they require to be limited to tasks in which there is a genuine common interest.

The economic implications of greater world unity are also serious, but to a great extent the poverty of Asia and Africa compared to America and Europe raises problems whatever political developments take place, or whatever foreign policy we follow. Within each country the State representing the people takes some responsibility for the poor. It goes further in most countries and assists particular industries (e.g. agriculture) or particular districts (e.g. 'distressed areas'); it does not merely support the poor until they can find work somewhere, it tries to bring work to them. The poorer countries are beginning to make the same sort of claim on the richer countries which poorer people—or poorer districts—make, say in Britain, on their richer citizens. The richer countries, notably America, have accepted this claim, and if some of their aid has been given for selfish reasons, why is not some of the aid given to our native poor given for selfish reasons, too? But it is a vast effort to try to raise living standards all over the poorer countries of the world. How are we to set about it? The greatest example of a country transformed from primitive poverty to the height of capitalist well-being is America. It was done by rip-roaring private enterprise and free immigration, not by any sort of 'Marshall Aid'. This being so, it is difficult to say that private enterprise cannot find the capital where the investment is worth while. Should we not then simply aim at *laissez-faire* and *laissez-passer*, encourage the 'under-developed' countries to maintain sound Governments and assure them that if they play according to these rules and really have resources to develop, they will get the capital all right?

I do not think we can. But rough though American progress was on the American Indian I do not think either that we can ignore its lessons entirely. What should then be our attitude?

First, I think that we must impress on everyone who wants to borrow that stable government is a prerequisite of a lender who wants to see results. Secondly, we must ourselves set an example in freeing trade and discouraging prestige expenditure on industries which are fundamentally unpaying. Thirdly, just as America kept out of Europe's wars during her early history, so we must accept that poor countries are natural neutrals. We must not expect them to take sides in the rivalries of the Great Powers. But even if we got agreement so far this leaves the bigger questions unresolved. Are India, Africa, the Arab States, to be encouraged to take a bigger share in world management while drawing more and more wealth off Europe and America? Or is there to be no representation without taxation? Are we really to support Malta at a standard of living equal to our own? Are we to be committed to finding help for ever-growing populations of little skill at the expense of developing our own highly skilled workers?

I emphasize that those who like myself want to see a Pluralist International Society may make the problem more difficult. But we do not create the problem. I think we have to reserve some rights over our own civilization, our own achievements: that is why I do not see a World Government coming into being in the meantime. I believe we have to bring in the representatives of the underdeveloped countries to take the unpopular as well as the popular decisions. As I write, India is asking the Western world to extricate her from the consequences of her own too hasty attempt at industrial expansion. She may claim that unless she rides the bicycle of progress fast she will fall off into anarchy or Communism. Her difficulty is a genuine one, her aims are praiseworthy, but nevertheless though we may well both as her old tutor and modern neighbour take some responsibility for her troubles, the main responsibility is hers. We have the right and indeed the duty to say to her statesmen that if the people of New York and Milwaukee, London and Edinburgh and Toronto are going to be told to give up some of their earnings to bail out an Indian Government heavily in debt, the Indian Government itself must explain to Indians how the situation arose and must itself, if India cannot stay on the bicycle without being a menace on the road, find a new means of progression. If this entails some regimentation we should not let it deter us from giving as much as we can afford and India can use.

One of the main reasons for wanting the Free World to be an area of economic freedom and high investment is that it should be able to help less fortunate countries. In the chapter on 'Economics' I have mentioned the importance of improving the financial backing for trade. I have also pointed out that the type of economic nationalism which leads to every country attempting to have every industry not only hampers the major industrial countries' development, but will be fatal to those countries which cannot possibly support such a programme. I believe that Liberals must accept a partnership between public and private capital in such areas as Africa, Asia and South America. I would like to see public capital directed more to the provision of such things as roads, universities and housing and private capital concentrated on industrial development. But the improvement of the world's distressed areas will ultimately depend on the response made by the people of these areas themselves and that, in turn, depends upon the aims which they can be given. What can be done among free peoples if the right aims and the right means are found is shown by the German (and, indeed, the Italian) recovery since the war. Here one potent means of releasing such energies has proved to be 'The Idea of Europe'. This is a thoroughly Liberal approach. While we advance towards a more Liberal world by the initial stage of closer collaboration among people who think alike and have common interests we must keep a constant goal in our mind's eye, even though that goal may be at present unattainable and some parts of the way towards it may still have to be explored. We have to clear away many of the obstacles to freedom, opportunity, peace and prosperity, the accretions of mistaken philosophies which, like barnacles on the hull of a ship, have retarded progress. But we must have no wavering in our determination to go to the end. Our aim is to create a Liberal pluralist world outside the Iron Curtain and eventually to capture by peaceful means the allegiance to humanism of the Communist countries themselves.

There are two subjects which though not usually treated as part of international or foreign affairs are nowadays closely related to our position in the world; one is Human Rights and the other is Defence.

The various declarations about the rights of individuals have always been treated by the countries involved as aspirations only. America has never taken them very seriously when it comes to Little Rock nor Britain for that matter in some of her colonies. That is not to say that America or Britain are not moving steadily towards equality of rights in most of their territories, but they are doing so because they recognize that they ought to do so and not, primarily, because of any international agreement. The declarations and agreements indeed are not only aspirations but aspirations by sovereign states. It is left to states to accept the declarations and it is left to them to do what they can about enforcing them.

Rights are of peculiar interest to people who are not in command. It is the natives in South Africa, the negroes in America, the Protestants in Spain, Roman Catholics in Northern Ireland and such like who are in greater or lesser need of some protection for their human rights. And the sovereign democratic nation-state has difficulty in knowing how to secure them their rights or indeed in knowing what these rights should be, once there is any real clash of interest. For if you run the State on the British system under which a parliamentary majority form the Government, and can push through their policies virtually unchecked, the minority have no guarantee of protection. The difficulty of modern democracy is seen in Cyprus. If you plant in Cyprus a parliamentary system on the British model the Turkish Cypriots will be permanently outvoted. That our past policies have inflamed the tempers to a heat which they need never have reached is beside the present point. The point is that tempers are liable to be heated when minorities are involved as Czechoslovakia, Ireland and Algeria, to mention only a few examples, have shown all too clearly. In Britain we have not faced a serious difficulty over a minority for some time. We have some give and take in our politics which makes it possible for minorities to exert influence. Our law protects human rights. But let no one think that we are beyond reproach. Our electoral system denies all representation to Scottish and Welsh Nationalists while it grossly under-represents the Liberal Party. The Conservative and Labour Parties exclude Scottish and Welsh Nationalists from official political broadcasting. There is already some feeling against West Indian immigrants, which shows that we are far from immune from colour prejudice.

Liberals must make clear that they believe in granting rights to take some part in government, rights to equality before the law, rights of opportunity in education and the professions, rights to free association and worship to everyone. They stand for the abolition of all distinctions in a state on grounds of colour or religion. They believe that minorities must not only share in these rights, but a place must be found for them in running society and government which means, again, experimenting with new forms of democracy. At present progress towards the ideal in most cases can best be made by leaving it to states themselves under the influence of informed world opinion. But world opinion must be expressed. Indeed, just as the fight against Communist tyranny is a world fight, though carried on by peaceful means, so the fight for human rights crosses the boundaries of states. What happens in Algeria is of concern to British Liberals. The motto about minding your own business has led to a great deal of avoidable suffering. Many great human reforms were made by men and women who resolutely refused to mind their own business. The treatment of people in Africa or Asia by European countries is our business, not only morally in the sense that we are our brother's keeper, but as a practical political matter because what one national power does reacts upon the others. There are some parts of the world, and Africa is one, where it is highly desirable that such powers should conform to a common code. The idea of trusteeship which is often said to lie at the root of British colonial policy and which was explicit under the mandate system of the League implies that the trustees' powers are not absolute.

Here again, however, we have to deal with things-as-they-are and they are neither simple nor uniform. Ultimately, most minority problems in a country where there is not a fundamental difference of race and outlook are resolvable by suitable political institutions and by the devolution of power. Conversely they will be insoluble if unbounded political power lies in the State and the machinery of the State's Government is in the hands of one party. A written constitution which limits the power of a Government is not to be wholly despised as an instrument for protecting individuals and minorities. Another method is to develop political institutions which do not entirely depend on counting heads. 'The sense of the meeting' can take into account dissentient opinions in a way decision by voting cannot. There are places, of which Ireland in the long run appears to be one, where because the minority are self-contained in an area which can be delimited, federation is the solution.

But we are still left with the cases in which there is a sharp division of colour, race and civilization: the cases which should be eventually solvable by Liberal principles and by new democratic powers, but which in the short run present formidable difficulties not to be met entirely by sweet reason. There are peculiarly the cases where a white minority lives in an African country. In such cases Liberals should make it clear that in the long run they believe that the native majority must take over power, but that for five, ten, twenty years or whatever a suitable time may be the predominant colonial power should run the country while the elements in it learn to live together and govern themselves. A time limit should be set to this period of tutelage whenever possible. And it should be long enough in difficult cases to allow those of the minority who cannot accept the ultimate transference of power either to leave or at least to draw together into well-defined districts. Until the time limit is reached the tutelary power should be firm and govern with strength.

As to questions of immigration: I hope that a supra-national authority will eventually be set up to control these. In the meantime *laissez-aller* is a principle which nearly always benefits communities as well as individuals. More countries have benefited by minorities than we usually remember. The influx of Jews into Europe has been a great boon to us and those countries who have expelled them have been the poorer for it. America would not have reached her present position of power and prosperity had she not been open to poor immigrants from despised races. So while I would concede that in the existing state of the world people of one race living together have some theoretical right to limit immigration, I know of no practical case at present for using this right.

Finally, there is defence. The mechanics of defence are not a matter which can be settled by Liberal principles. They are outside the scope of this book. What does concern Liberals is that overall defence of the Free World is clearly something far beyond any one country. It is for that reason that the Liberal Party has worked to make N.A.T.O. as strong as possible. It is their policy, too, that weapons which could only be used in a world war, such as nuclear bombs and rockets, should be controlled by an international authority. This has led to misunderstanding. The Liberal Party is not a pacifist party, and as I have said earlier in this chapter Liberals must be prepared to defend their principles. Nor is it relevant to accuse Liberals of sheltering behind the Americans.

The Americans as much as anyone else must forego their 'right' to plunge the free world into nuclear war, and for them as much as for us defence is global. No nation can contain a world war once it has broken out, nor contract out of it before it does. If they are in the way they will get hurt as Belgium found in 1914 and Holland in 1940. The sensible defence policy then is for the free nations to combine together and for those among them who are industrially strong to make a contribution to nuclear as to other forms of defence. As America has far more nuclear power than any of her allies the nuclear programme as a matter of convenience will centre round her effort. If she does not want contributions from her allies, well and good, but it does not relieve them or her from agreement about the circumstances in which nuclear weapons should be used.

There is a danger at present that many nations will soon have weapons capable of starting a world war and gravely damaging civilization unless a pact is formed by the principal free nations to enforce international control. There is a further danger that if each nation prepares itself for nuclear war, some nation will gear up its defence system so that once certain buttons are pressed the whole machinery goes into operation. Unless it is that once the armed forces are pushed down 'the pipe line' it is impossible to limit them, it is difficult to understand the massive flow of American might, including nuclear weapons, into Lebanon. The amount was out of all proportion to their task and quite out of keeping with the reasons alleged for sending them there. Apart from control of nuclear weapons, there is then an urgent need for agreement about the circumstances in which they are going to be used. Presumably there are already commanders who have the authority, given some Russian attack, some identification of hostile rockets fired from the east or whatever it may be, to take counter-action with nuclear bombs or rockets. Once they have done so, it may not matter whether the counter-attack is launched from Texas, Norfolk or Normandy. The Russians will retaliate where it seems to them most effective and not merely on those bases from which the counter-attack came. And for what it is worth, so long as N.A.T.O is a close alliance and American forces are stationed in many different countries they will have right in international law to do so.

BRITISH LIBERALISM AND REST OF WORLD

Here, then, is a field in which sovereign states will be bent on suicide if they do not pool their sovereignty and pool it quickly.

The usual objections to such a policy are that unless Britain has nuclear weapons she will lose her influence and that she may be in a position where she has need of her own deterrent. The first objection which is made the reason for the current Labour Party policy, namely to make and hold these weapons, does not seem to me to carry much weight. The nuclear weapon is, if of any value, of value as an ultimate deterrent. It is not an instrument of diplomatic policy. Indeed, force of this order is out of date and self-defeating. No one considers that India or Western Germany are negligible countries which would be greatly strengthened by possessing nuclear weapons. By all means let us try to strike a bargain over nuclear disarmament but do not let us delude ourselves that being a nuclear power makes us much more respected by our ill-wishers. But what about our allies? Some people believe that we should be at the mercy of American diplomacy if we did not control our own nuclear weapons. It must be remembered that Liberals demand some control over all such weapons, especially American missiles. But even if we did not get it, the mind which envisages allied diplomacy depending on nuclear force at the disposal of each ally is scarcely living in a real world. If the Americans are only going to be deterred from folly by the relatively tiny nuclear armaments of Britain, then France, Germany and every other ally must demand to lay their bombs on the table to get a fair hearing. It is a picture of an inane mad-hatter's tea-party. How the possession of nuclear power is to be used to cajole or bully the Americans is not clear. We should presumably hardly launch a world war on our own, but to threaten it without counting the cost of our bluff would be extremely foolish.

The second objection is more worthy of respect. It is possible to envisage a situation in which we might feel that we must fight for some British or European interest but in which America would feel that the candle was not worth the game of world war. Russia might be deterred from some aggression in Europe by the suspicion that though she might feel certain that America would not react violently Britain would, and if Britain possessed nuclear missiles the deterrent would be effective. Some European countries may indeed welcome the fact that we have the bomb. So indeed may America, for she may not want the whole responsibility for exercising the deterrent to fall on her shoulders.

I concede that the Western alliance might fall into such a tragic state that even the limited amount of interdependence so far achieved could have been disrupted. Certainly the Western Alliance would have become very different from what Liberals want it to be. Secondly, the force of the deterrent depends on the will to use it. A British deterrent will only deter if it is clear that the British are willing to start a nuclear war from which America stands aside. Of course, we may expect or hope that she will be dragged in. But to rest one's policy on such a line of thought is disastrous. Lastly, would our deterrent deter?

Apart from the heavy burden on our economy of keeping up our own deterrent, on grounds of foreign policy as well as defence, it is surely more sensible to try to make interdependence real through a common cause than to build our own nuclear deterrents when one overall deterrent is enough. But Britain should then try to reach some clearer agreement with her allies. She should try, too, to clear her own mind about the use of force in the modern world.

I do not believe that world war is likely to break out by mistake, though it is a possibility. It is, however, by no means impossible that circumstances may build up in which the leashes on the dogs of war wear very thin. One of the clearest cases of a war of deliberate aggression was the attack of America on Spain. Pacific democracies let alone autocracies can work themselves into a state in which war is almost inevitable. There is an element of the mad dog in many Governments; the temptation to cut Gordian knots: the intoxication of appeals to honour. The forces of sanity, the will of the people itself can very easily be stampeded at a certain moment. It is the curb on the build-up of a war situation which must be strengthened, and strengthened by much clearer recognition that force must be repudiated as an instrument of policy. The circumstances in which ultimate normal force becomes justifiable must be circumscribed as closely as possible.

This means the strengthening of international order which has already been discussed. But it also means the retention of some conventional forces for tasks not yet assumed by an international police force, and for the maintenance of order in territories for which Britain is responsible.

But here, again, the situations in which intervention by conventional forces is justified need some definition. Until a U.N. force can take over it is desirable that sufficient conventional forces should remain in Europe to prevent a Russian coup across the frontier. It is desirable that in neutralized areas of the world, such as the Middle East might well become, some 'police force' should be stationed: though in this case it is difficult to see how it could be other than international, unless supplied by the countries themselves. The obsession with indirect aggression is dangerous. To begin with, as I have said, Liberals cannot forego the effort to try to convert Communists, nor the attempt to encourage Liberal elements in other countries. This should certainly be done by peaceful means but it may ultimately lead to revolt. Indirect aggression is usually used to cover activities going further than mere propaganda: it includes the sending into other countries of saboteurs, the use of bribery, the running of arms to rebels and deliberate campaigns of threats and economic war. But it will only be effective where the Government against which it is directed is unpopular or grossly incompetent. Armed intervention by conventional forces such as we have seen in the Middle East is not either justifiable or effective against such indirect aggression. It creates more problems than it solves: it may lead to just the sort of atmosphere in which war spreads: and it diverts our attention from trying to establish an effective policy to remove the causes which make indirect aggression pay.

Trade, economic and financial development, the guarantee of human rights and defence are some of the subjects which need new political ideas and institutions which will cross State boundaries. For the safety of men and for their prosperity it is urgent that we devise new politics. A new political system must be more than the stylization of the struggle for power. Politics must answer to man's purpose. The world-wide political problem today is to devise democratic systems which will be sufficiently free from corruption and capable of organizing material advance. Parliamentary government is not the universal solution. Election of an autocratic executive is as far as democracy can advance in some countries at present.

10

Epilogue

The picture I have tried to draw in this book is of a Liberal world in which everyone has a wide range of choice and opportunity, a world in which the variety of things to choose from is great and in which there is an opportunity to choose things of good quality and not merely in great quantity. That means restoring to individuals much of the power which has been filched from them. It means leaving them with more money to spend after they have paid their taxes. It means giving them the opportunity to take a more active part in running their own lives. It means giving them the chance to take part in some bodies through which they can run activities which particularly interest them. For a Liberal it is not enough that everyone should possess some vague residual power by vote at a general election; we believe that while democracy does not mean that every governmental activity must be supervised by a supreme legislature and executive responsible to everybody, it does mean that those who are particularly interested in particular activities should have an opportunity of taking a more effective part in them.

The role of government, and by government in this context I mean all the bodies which Liberals hope will share governmental power in a Pluralist Society, is twofold. It must ensure that the Liberal Society stays Liberal by maintaining the conditions under which it can work. Secondly, it must from time to time intervene to undertake those services which are properly government services because they are necessary to a whole group of people who would nevertheless find it difficult to supply them for themselves.

I must stress again that I do not think it enough to remove the obstacles to Liberalism. There is a school of thought which seems

EPILOGUE

to believe that if you free trade, reduce interference by the Government, and let the market take its course Liberalism will flourish. I hope that I have made it clear that I do not belong to this school. And the main reason why I do not belong to it is that it seems to assume that a Liberal world is in some sense more 'natural' than any other. But the free market—to take that as an example of a Liberal institution—is a highly 'artificial' arrangement. It depends at least on orderly government, impartial laws fairly enforced and a currency which is to be trusted. The same is true of democracy itself as practised in Britain or America. More important still the world is never neutral. I would certainly agree with what I would call the negative school of Liberals, that if Liberal principles are taught and if they are seen to be practised, they will win adherence. I believe they are better and that they work better than say Socialist principles, and that the majority of people will recognize this if they are given a chance. But they must be given the chance. The individual upon whose energy and judgment Liberals rely must be taught to appreciate Liberalism. By that I do not mean, though, that he is to be brought up on Liberal Party propaganda to the exclusion of everything else. What I mean is that he must be taught to use his mind, he must be shown the implications of various ways of life and he must be taught the importance of logical thought and tolerance of other people's points of view.

While I think it is very important to remove the obstacles to Liberalism, I would go further in the field of government and say that a Liberal Government must not be frightened to take positive action to promote Liberal conditions. Defence of the weak, help for the poor, eradication of evils, guardianship of scarce resources, maintenance of the coherence of society, the setting of an example in those fields where the Government acts, all these are part of its legitimate duties.

As this is not a party programme, and as it is of the essence of Liberalism that methods must be changed to meet new needs—Government is made for man and must change as circumstances change—I do not feel it necessary to cover every subject or to give detailed suggestions on those subjects with which I have dealt. It is the main outline of a Liberal society as they affect the individual and his Government that I have tried to draw. If we are to get a

EPILOGUE

Liberal society we must aim high. We must start by defining our aims.

The Liberal Party today must be thrusting, progressive, enterprising in outlook, indeed it must always be progressive for it can never be satisfied with less than a fully Liberal society. Such a society will always need to be continually reconsidered: the roles of the State, the Government and the Nation need to be recast from generation to generation. But the ideal remains. In having its ideal of the Free Society Liberalism differs from both Conservatism and Socialism. For the Conservative is content to inherit what history has given him and make the best of it. He has always hankered after a closed society. He has not rebelled against the Socialist philosophy of the State, indeed he finds it in many ways congenial to his eighteenth-century, restrictionist and hierarchical frame of mind. He shares the Socialist's pessimism about the ordinary man: the Tory fears the inevitable triumph of original sin unless the great majority will submit themselves to the leadership of an *élite*—in conservative dreams, a hereditary officer-class conservative *élite*. The Tory, like the Socialist, relies a great deal on vague emotive words to conjure up the feeling of Hegelian inevitability and to save himself from having to take a rational initiative. The Socialist, too, rejects the Liberal ideal of a Free Society as impractical. He, too, is obsessed by original sin: though he accepts the need to change, progress, improve, his progress is towards an all-powerful state and away from the devolution of power to individuals.

In some ways, indeed, the Socialist like the Tory seeks to evade facing the difficulties of a changing world. He wants to clamp the living society of mankind into a Socialist vice. But the world will break out eventually from any vice. Burke, with all his greatness, underestimated the sudden jolts of history. History is very uneven. Liberals recognize this: they want to study, direct, engineer change towards maintaining the Liberal society. They hope to make it fruitful and peaceful. They do not pretend that they can stop it.

Liberalism is a difficult creed, for Liberals must convince and not merely command. The Tory has no urge to experiment and explore. He needs no ideal, for he is content to conserve whatever

EPILOGUE

is to hand. He is the hermit crab of politics sheltering in whatever shell, Liberal or Socialist, lies to hand. The Socialist is prepared to force his Utopia down our throats. It is always easier to use force than reason. It is always more difficult to appeal to man's reason and better judgment than to his conservatism or his selfishness. But the more difficult appeal must be made. There can be no free or peaceful world unless it succeeds.

Appendix

Again I repeat that this book does not pretend to give an immediate political programme. But it may be convenient if I finish it by adding a summary of Liberal aims, as I see them, in the world of the fairly immediate future; a future which may be less remote than we now suspect.

We hear a good deal about Opportunity but not so much about Responsibility, which is the reverse of the coin. Personal, political and economic responsibility are all weak today. And so is leadership. Too many people hug committees because they offer release from individual responsibility. Too many people hope to throw their troubles on to someone else. Responsibility is not popular. All the more reason for so organizing the affairs of the Western World that we make the most profitable use of those who will assume it.

The gulf between 'us' and 'them' is wide: it divides the governed from the Government, the worker from the manager, the citizen from the Civil Servant.

While in some directions opportunity is widening, in the most exciting directions of all, those offering opportunity to work together for a common aim, it is diminishing.

The present situation is well illustrated in the title of the latest policy statement by the Labour Party, 'The Future Labour Offers You'. The implication is that 'you' are different from the Labour Party. All you have to do is to sit tight, except on polling day when you must vote Labour. After that you can relax and gather in the benefits.

These are the politics of a proletariat.

POLITICAL REFORM

Liberals, therefore, set store on *political reform* designed to give

APPENDIX

everyone more opportunity and to set aims before them which will justify the responsibility opportunity entails: such developments will also enable us to handle our joint affairs more successfully. Before ever Parliament meets, the shape of our politics is determined by our political and social system. This system needs reform.

At present the single-member constituency, plus parties largely dependent on organized employers and organized Labour, with no counterbalance by an effective second chamber, means that pressure groups have too much weight: independent and able men are reluctant to enter politics, conformity is at a premium. Further, the pressures on M.P.s and the Government generated by sectional interests, and making for ever-increasing government interference and expenditure, are magnified at the expense of wider views. There are also some activities, such as the supervision of the Nationalized Industries—for which most members are not only unfitted but on which there is a lack of those pressures which are the final guarantee of attention to the public interest. For most members there is no great incentive to unravel the finance of these industries nor to consider the implications on the gilt-edged market of finding money for capital schemes of unproven utility.

First, then, let either electoral reform be introduced for the House of Commons or the members of a second chamber be chosen by indirect or proportional election.

Secondly, let contracting 'in' take the place of contracting 'out' of the political levy, and ensure that any subscriptions to party funds from such bodies as limited companies, are made only with the consent of their shareholders.

Many appointments and much patronage should be taken out of the hands of the Government and either entrusted to the bodies concerned (e.g. the Church of England) or made or exercised on the advice of a committee of the Privy Council or other independent authority. Moreover, the extent of patronage should be reduced.

Steps should be taken to ensure that the channels of information and opinion are neither twisted nor blocked. Control of political broadcasting should be left to the B.B.C. or I.T.A. under general directions to hold a balance but to allow all views access to radio and television. If necessary the independence of the Press

APPENDIX

must be guaranteed by positive steps such as subsidizing newsprint or setting up a stronger Press Council.

New methods must be found of controlling such industries and activities as must be run publicly. Parliament, our heritage from Simon de Montfort, remarkable as it is, is not the only possible channel for the Public Will.

By these reforms we should be able to improve the standard of participation by the people at large. But we still need reform of our political institutions themselves.

A political leader can now speak direct to the electorate by television. This weakens the importance of the private member as a link between citizen and government on broad political questions. The House of Commons is often by-passed when important matters are dealt with by direct consultation between the Minister and the interests concerned. The House of Commons' chief roles are: to supply a Government—or most of it; to control it and to amend its legislation; to serve as a place for public discussion; to serve as a place where grievances can be aired.

It is essential to take some of the burden of work off M.P.s entirely and to arrange the rest so that it can be more effectively discharged. Therefore, let us allow certain business to devolve on Scottish, Welsh and English Parliaments, and delegate the control of such industries as may remain nationalized to those directly concerned, subject to competition and ultimate governmental supervision, with assistance of expert bodies such as an investment board. It is high time, too, that we redrafted the areas and functions of local authorities and gave development functions over to authorities largely controlled by those interested (in all senses) in these functions.

By these reforms more people can take responsibility for running the country, by becoming directly involved in the control of functions about which they are knowledgeable or in which they have an interest. They will also have opportunity.

As for Parliament itself, it too needs reform. The second chamber should be strengthened as a deliberative, controlling and amending chamber. To achieve this it must be freed from its hereditary element and chosen either by proportional representation or indirect election.

APPENDIX

The House of Commons should experiment with standing committees on various topics, e.g. the Colonies and Foreign Affairs or Expenditure. Members of such committees must be adequately equipped with staff.

No Member should be allowed to propose extra expenditure without indicating how it is to be met. Some services should be financed by taxes ear-marked for special purposes.

Defeat of a Government in committee or, indeed, on votes other than those of confidence in the House itself, should not be regarded as a humiliation hardly to be borne without resigning.

Government business should not be allowed to edge back into the space cleared in parliamentary time by the above reforms. This space should be used for debates on major motions proposed by private members and for enabling a reduction to be made in the overall total of sitting hours.

Private members should be better equipped, with secretarial assistance etc., to handle personal matters of constituents. They should also have ready access to lawyers, accountants and other experts who would relieve them of much routine work perhaps better done by others, and leave them more time for efficient supervision of such questions as, for instance, government expenditure.

TWIN SIDES OF LIBERAL POLITICAL ECONOMY

A Liberal economy will have two sides. On the one hand it will ever extend the area of choice and opportunity for the individual by sustaining a free-competition economy. On the other hand, it will use the free competitive economy to support certain common aims which will give people something to work for together.

To improve the free competitive economy we need lower taxation on earnings, more effective Parliamentary and Treasury control over expenditure and reform of the financing of nationalized industries. We must put such industries in a position where they can raise capital (apart from that portion which they can supply themselves) on the market. An investment board for considering their needs should give valuable guidance if properly utilized. At present the enormous capital demands, for which inadequate justification is offered, apart from distorting the economy in other

APPENDIX

ways, seriously disrupts the long-term gilt-edged market. We need, too, currency and credit reform, internationally and at home. For instance, the reserves of the world must be increased by increasing the funds at the disposal of the International Monetary Fund or by raising the price of gold: indeed, the world must set itself free eventually from its present too great dependence on gold. We should put an end to the automatic connection between credit and government borrowing, and we should devise new methods of credit control, e.g. by changes in ratios.

To safeguard the public through competition we should strengthen the Monopolies Commission and the Restrictive Practices' Court, enabling the former to review such industrial changes as take-over bids tending to monopoly.

To get expansion without inflation we must spread ownership and profits much more widely. This process can be encouraged by tax reform which should be coupled with a reform of the Company Laws to give status to workers. To derive the maximum from a competitive system we need to envisage large markets. This means breaking down trade and political barriers so that industry may be organized in big units without necessarily becoming rigid and monopolistic.

The difficulty over the balance of payments which has dogged this country since the war is to a large extent a reflection of the inadequacy of liquid reserves combined with our failure to devise a financial system which will allow of our resources being fully employed without inflation. This in turn is connected with our failure to achieve as full productivity as should be possible and our inefficient and antiquated method of settling wages and raising revenue from taxation.

It is possible to set down under six heads the reforms we envisage; they comprise:

Social Services which should eventually become a basic income for all. In the meantime, basic pensions should be paid out of a social services tax. Those who want a pension above the basic minimum should be offered voluntary schemes, the Government itself operating a voluntary scheme for those not otherwise covered. Unemployment pay should be the life-line by which those who are flung out of work by technical change etc. climb back

APPENDIX

into useful employment. It should allow generous lump-sum payments and should be coupled to facilities for re-training.

Our slum areas must be rebuilt in the name not only of health and comfort but of beauty. The urban shambles must be checked and the countryside reborn by a policy of assisting agriculture to be efficient, of spreading industry to smaller towns, and of revitalizing those areas, such as the Highlands of Scotland, whose populations are draining away for lack of employment.

We should offer improved 'secondary' education and University education, not only to an increasing number of our children, but to those who want to come here from overseas.

As to the world at large, we should accept the obligations to help the poor wherever they are, but since our resources are limited, let us start in those places, such as Europe, India and our African colonies, with which our destinies have already mingled. Help should be given not for strategic or selfish reasons but because we owe it to other, less fortunate, human beings.

DEFENCE

The 'H' bomb may not alter the moral aspects of defence, but it certainly alters the practical possibilities. A defence policy must now be different in kind from its predecessors. There is no defence against total nuclear war—only deterrence. Deterrence cannot be 'national'; so long as we rely on nuclear deterrents, they must be Western deterrents. The building of a British deterrent is the sign of out-of-date thinking about Foreign and Defence policy and is wasteful as well as futile.

To my mind there is no valid distinction between 'tactical' and 'strategic' nuclear weapons. Until we get comprehensive disarmament, which is the ultimate aim, we and our allies need a joint policy of deterrence, supported by well-equipped, mobile conventional forces.

WORLD AFFAIRS

The Western World must get its mind away from late nineteenth century political thinking. It must be willing to explore in politics as in science. It must learn from the Communists: in particular it

APPENDIX

should study their success in identifying their people with the aims of their governments and their handling of multi-racial countries.

A tragic feature of the last few years has been the decline of British leadership. Under a crust of unsure jingoism there is much defeatism among the British themselves. Certainly we should not rely on force to keep our end up in the world, but equally, if we are going to claim to be the Greeks of the Modern World we must behave like Athenians. It is not enough to keep on repeating that we have so much experience of handling world affairs that we are, without question, more skilled than other countries. Of course, we have much experience; we have much expertise, but we still cannot make the mistakes we have made, for example in Cyprus. We must define our purposes and give much thought as to how they are to be achieved.

High on the list of these purposes I place Liberalism. One facet of Liberal belief is that people should be allowed, and encouraged, to shape their own lives, though the methods by which they achieve this shaping, with due respect for the rights and aspirations of others, will naturally differ. Under this heading of Liberalism I include not only financial and technical assistance to other countries, but political experiment of many kinds. We should assist new groupings among the emergent countries, e.g. in the Middle East. We must not lead people to suppose that democratic advance necessarily means the creation of a nation-state and a Parliament on the British model. In many countries the American system of electing a President and endowing him with considerable powers, may be more appropriate. We should certainly encourage federations and confederations of states in Africa and the Middle East. We should concentrate on devising practical means of pushing through economic development and on the training of executives and civil servants in our colonies.

Another purpose of foreign policy is Peace, which entails more than alignment against Communism. Peace demands continual negotiation, on Western initiative, with the Communist world. It involves, too, recognition of the desire of many countries for neutrality. The field where it is probably most urgent for the Western Alliance to pool the sovereignties of its component states is that of foreign policy and defence.

APPENDIX

And with peace must go the surrender of some sovereignty and the building up of means for peaceful change supported by our international police or observer force. Britain can fruitfully start on this work in her own Commonwealth and in Europe as well as through the U.N.

In the Commonwealth we should aim at a Commonwealth service recruited for technical, economic and educational work throughout the Commonwealth, and indeed in such countries outside as may ask for its help.

We need also a Commonwealth Council to deal particularly with foreign and defence matters, and a Commonwealth Bank. In time we may hope that there will be Commonwealth armed forces. The Privy Council might be used as an instrument for guiding some of these experiments.

In Europe we should make the Committee of Foreign Ministers a reality and join the Iron and Steel Community and Euratom. We should set about the creation of joint machinery for economic planning for Europe as a whole.

I am finishing this book at a time of more than usual bewilderment in political affairs. We are leaving behind the problems and policies which have formed the background of the last sixty or seventy years—those associated with nationalism and state socialism. In the foreground we shall soon see the retirement of the Father figures of the West, President Eisenhower, Dr. Adenauer and General de Gaulle. The groupings under the political kaleidoscope will soon be re-shuffled throughout the free world—and probably, owing to the rise of China, within the Communist world as well.

It is apparent that Communism is going to dominate part of the world political picture. What is going to dominate the rest? In any free political structure there will be Conservatives. But what is going to be the political dynamic which should balance Conservatism in a democratic system? To say that it is going to be a Liberal Party or Parties is not enough, unless Liberalism is re-written in modern terms. For one thing, Liberalism cannot afford to be negative. It must give scope for unselfish aims. For another, it cannot afford to be disruptive. The best non-conformity has always been co-operative, such co-operation being volun-

APPENDIX

tary and not imposed. Thirdly, it is incumbent on Liberals to show that the essential liberal aims are compatible with movement towards a higher standard of life in the underdeveloped countries. Economic progress and political freedom must not become alternatives. This means that Liberals must be ready to modify the free enterprise system to allow of rapid investment and mobilization of resources, the prerequisite being that we must gain the voluntary and active co-operation of the people concerned. Fourthly, Liberalism must meet the problems of multi-racial societies, and those arising from the need for different political groupings, by breaking down the sovereignty of the nation state. Apart from the desirability of these things for their own sakes, the might of China is ever at our backs.

If a British Liberal Party can sustain such a programme it has a bright future as the progressive alternative in politics. It will not, however, be achieved without rebellion against restriction within our present political and economic framework. Unless we can break out of the bonds which constrain us and mobilize more people for political enterprise we are doomed to such narrow margins that we shall always remain the slave of events. By which I mean, for instance, that in foreign affairs we shall always be kept on the defensive because of the difficulties of reaching a common policy between theoretically sovereign independent governments: in economic affairs we shall be continually on the knife-edge between inflation and stagnation. The result of these compulsions can be seen in the growing together of Conservative and Labour policies. They are becoming as similar as American cars and each is forced to vie with the other in vote-getting embellishments to a basically similar design.

In the next few years a new vision and vitality must invigorate the politics of this country. Since, with all our mistakes, we British remain the most deeply political of people, if we allow the springs of democracy to run dry there is no telling what may happen in the West as a whole. There is no room in Britain today for a third party working in the same context as the Tory and Labour Parties. The need is for a Party outside that context, seeing the prizes rather than the difficulties, calling for effort rather than perquisites, and asserting those issues, such as the

APPENDIX

international issues, which do not at once impinge on many people's imagination because they are as yet not immediately felt in the pocket. Nevertheless, the party may find itself in a situation where its support is essential to enable either the Conservative or Labour Parties to govern. It must then so conduct itself in Parliament as to make stable government possible, and it would have the right to expect the majority party, whether Tory or Labour, to make concessions to the same end. Obvious concessions at the time of writing would be for the Tory Party to abandon 'Suez' type diplomacy or for the Labour Party to forego further nationalization. At the same time a Liberal Party would have to try to give Tory and Labour governments the courage of their more Liberal convictions, for instance, on trade and law reform.

But such short-term parliamentary actions would not deflect the Party from the pursuit of its major and urgent aims for the future of politics at large.

Index

Advisory bodies, 75
Africa and Asia, 162, 169, 174
Agriculture, 76 et seq.
Algeria, 174
America, 17, 43, 47, 70, 71, 121, 175, 176, 177
Arts, the, 131 et seq.
Arts Council, the, 136, 137
Asia (*see* Africa), 169
Asquith, Mr. H. H. (Earl of Oxford and Asquith), 27, 98
Australia, 131
Austria, 155
Aviation industry, 56, 57

Balance of Payments, 70
B.B.C., 147
Belloc, Mr. H., 89
Beveridge, Lord, 28, 68, 105
Bismarck, 99
Boarding schools, 120
Bosanquet, B., 52
British Commonwealth, 136 et seq., 163
Broadcasting, 39, 147 et seq.
Brussels Treaty, the, 167
Budget, the, 69
Burke, 13, 19

Catholic Education, 127
Chamberlain, Mr. Neville, 24
China, 155, 167
Christian Socialists, 26
Christie, Mr. John, 137
Choice, 13 et seq.
Churchill, Sir Winston, 25, 40, 42, 164
Clark, Mr. Colin, 67
Cohen Committee, 41, 42
Competition, 54 et seq.

Communists and Communism, 19, 70, 151, 160 et seq.
Co-ownership, 79 et seq.
Copeman, Mr. George, 83
Covenant of the League of Nations, 168
Covent Garden, 137
Criticism, 12
Crofters Commission, 78
Cyprus, 46, 173

Darwin, Charles, 24
Death Duties, 149
Defence, 36, 172, 175 et seq.
Democracy, 36
Distribution, 89, 90
Division of Labour, the, 74, 96

Edinburgh Corporation, 136
Education, 104, 113 et seq.
Electoral reform, 40

Family Allowances, 109
Fatstock Marketing Corporation, 77
Foreign Policy, 151 et seq.
Free Enterprise, 59 et seq.
Free Trade, 72 et seq., 157
Free Trade Area, 165
Freedom of Speech, 12
French, the (and France), 23, 177
Friedman, Professor Milton, 104
Further Education, 126

G.A.T.T., 73
Germans and Germany, 21, 25, 177
Ghana, 162

Health Service, 107
Hegel, 24 et seq.
Herbert, Sir Alan, 40
Highland Development Authority, 76

INDEX

Highlands, the, 77, 141
Hire Purchase, 88 et seq.
Hitler, 19
Holists and Holism, 25 et seq.
Hollis, Mr. Christopher, 47
House of Commons, 34, 44
Housing, 105
Human Rights, 172

Immigration, 175
Independent Schools, 119
India, 169, 171
Inflation, 62, 67, 68
International Monetary Fund, the, 71
Investment, 57, 171
Ireland, 174
Israel, 60, 166
I.T.A., 147

Jordan, 46, 167

Keynes, Lord, 69

Labour Party, 21, 39, 54 et seq., 90
Land Tenure, 77
Laver, Mr., 100
Lebanon, 46, 166, 167
Lewis, John, Partnership, 86, 93, 137
Liberals and Liberalism, 19, 20 et seq., 27, 99, 105, 151 et seq., 159, 182
Lippman, Mr. Walter, 16, 24
Literature, 135 et seq.
Lloyd George, Mr. (Earl Lloyd George), 27
Local Government, 50 et seq., 141 et seq.
Lord's Day Observance Society, 37

Macmillan, Mr. H. (Prime Minister), 156, 162
Madariaga de, Senor, 110
Manchester Guardian, The, 84, 146
Marx, Karl, 24 et seq.
Mill, J. S., 50, 121
Monopoly, 66
Monopolies Commission, 66

Nasser, Colonel, 162
Nation, the, 23

Nation State, the, 22, 30, 154, 158 et seq., 169
National Assistance, 31
National Insurance Fund, 102, 108
Nationalism, 161, 162
Nationalized industries, 65, 76, 90, 91
N.A.T.O., 74, 156, 175
New Statesman and Nation, The, 146
Nuclear weapons, 176

Observer, The, 132, 146
Old Age Pensions, 27, 98
Opportunity and the Opportunity State, 58
Orwell, George, 18
Outrage, 132

Parliament, 41 et seq.
Peace and pacifism, 152, 153
Peacock, Professor A., 102
Pensions, 107, 112
Planning, 26, 32, 143
Popper, Professor, 24, 34
Press, the, 145 et seq.
Pressures and Pressure Groups, 45, 112
Press Council, 146, 147
Privy Council, the, 149
Productivity, 67
Profits (and losses), 26, 28, 30, 31, 58, 81, 86, 87
Proletariat, 17, 43
Property (private), 60, 79, 88

Quakers, 52

Rathbone, Miss E., 40
Race relations, 173, 174
Reason and the rational approach, 12, 13, 128
Responsibility, 12, 75, 80, 81
Restrictive Practices Court, 66
Ricardo, 26
Roads, 141
Rowntree, Mr. S., 106
Russia, 121, 155, 161, 176, 177
Russell, Bertrand (Earl Russell), 115

Sadler's Wells, 137

INDEX

Savings, 83, 91
School leaving age, 115
Scotland, 51
Second Chamber, 43
Shareholders, 80 et seq.
Smith, Adam, 19
Socialists and Socialism, 21 et seq., 24 et seq., 105, 151, 182
Society, 16 et seq.
Sovereignty, 47, 166
Standing Committees, 47
Sterling Area, 71, 74
Stock Exchange, 16, 84
Spectator, The, 146
Stigler, Professor G., 57
Suez, 12, 46, 167
Syndicalism, 91

Tatler, The, 15
Taxation, 67, 68, 72, 82, 92, 139, 140
Teachers, 129, 130
Technical Education, 151,
Time and Tide, 147
Tories and the Tory Party, 23, 34, 39, 45, 90, 120, 151, 182

Town and Country Planning, 69, 106, 128, 135, 138, 140, 142
Trades Unions, 16, 18, 27, 37, 38, 61, 62, 63
Transport, 46, 140, 141
Treasury control, 42, 69
Treasury Deposit Receipts, 69

United Nations, 153, 166 et seq.
United Nations' Charter, 167, 168
Universities, 124 et seq.

Wages, 61 et seq.
Wales, 51
Welfare Services, 27
Wolfenden Report, 149
Wootton, Barbara (Lady Wootton), 61
World Bank, the, 75

Yalta, 155
Yellow Book, the, 22

Zeiss, Carl, foundation, 61, 93

Date Due

JAN 16 1980	MAR 13 1996	
JAN 23 1980		
FEB 13 1980		
APR 24 1:36		
MAR 22 1999		
MAR 18 1999		
OCT 25 2006		
NOV 07 2006		

bdy CAT. NO. 23 233 PRINTED IN U.S.A.

JN1129 .L4G7
Grimond, Joseph
  The liberal future

DATE	ISSUED TO

180552

CPSIA information can be obtained
at www.ICGtesting.com
Printed in the USA
BVHW050006140223
658390BV00008B/250